A-Z
ANIMAL BEHAVIOUR

A-Z ANIMAL BEHAVIOUR

Prof. Manoranjan Tripathi

CENTRUM PRESS
NEW DELHI-110002 (INDIA)

CENTRUM PRESS
H.O.: 4360/4, Ansari Road, Daryaganj,
New Delhi-110 002 (India)
Ph.: 23278000, 23261597

B.O.: No. 1015, Ist Main Road, BSK IIIrd Stage
IIIrd Phase, IIIrd Block,
Bangalore - 560 085 (India)
Tel.: 080-41723429
Visit us at: www.centrumpress.com

A-Z Animal Behaviour

First Edition, 2009

ISBN 978-93-80106-01-4

PRINTED IN INDIA

Printed at Salasar Imaging Systems, Delhi-110035 (India)

Contents

Preface *vii*

1. Biology of Vertebrates 1
2. Animal Home Ranges 27
3. Habitat Evaluation in Delusions 65
4. Terrestrial Vertebrates 112
5. Population Density 132
6. Monitoring Populations 149
7. Population Viability Analysis 167
8. Mammalian Societies 209
9. Modeling Species Distribution 266

Index 307

Preface

The scientific study of Animal Behavior is known as Ethology. Ethology is a combination of laboratory and field science, with strong ties to certain other disciplines like Neuroanatomy, Ecology, Evolution. Ethologists are typically interested in a behavioral process rather than in a particular animal group and often study one type of behavior in a number of unrelated animals.

The desire to understand the animal world has made Ethology a rapidly growing field, and since the turn of the 21st century, many prior understandings related to diverse fields such as Animal Communication, Personal Symbolic name use, Animal emotions, Animal Culture and Learning, and even sexual conduct, long thought to be well understood, have been revolutionized, as have new fields such as Neuroethology.

Animal behaviorists have learned that often many answers are right in some respects, and almost all are wrong in some ways or in some situations. The important thing is not whether a particular answer is right or wrong, but rather the process by which new information is generated. In addition, each exercise requires, to an increasing extent, the students to play an active role in designing aspects of the exercise. This role shifts the students from passive spectators to active participants in the process of scientific investigation and gives them "hands-on" experience with real science.

Author

Chapter 1

Biology of Vertebrates

Vertebrates often are marked to facilitate identification of free-ranging individual animals or groups for studies of behaviour, population biology, and physiology. Marked animals provided data for many of the topics discussed in this volume, including home range use, resource selection, social behaviour, and population estimation.

Markers can be classified into three general categories: mutilations, tags and bands, and radiotransmitters. The appropriate marking technique for a study depends on several considerations, including study objectives, target species, marker cost, marker efficacy, and marker effects on the animals.

Studies using marked animals are characterized by the assumption that marking does not affect animals or that negative effects are not important. The assumption of no significant marking effects is critical because it is the basis for generalizing data collected from marked individuals to unmarked animals and populations. However, the assumption has not been tested rigorously for most marker types or animal species, despite the often necessary use of seemingly invasive marking techniques.

The general paucity of marker evaluation studies apparently is related to the difficulties associated with conducting such tests in the field, as well as the belief that marker evaluation is tangential to most study objectives and therefore of minor importance to the researcher.

In addition, studies that evaluate marker effects often suffer from small samples, thus leading to qualitative conclusions or weak statistical inference. Researchers tend to choose markers that intuitively seem least likely to induce abnormal behaviour or survival, even though data supporting that assertion usually are weak or lacking. However, if the assumption of no marking effects is violated and the effect is not evaluated, then data collected from marked animals will be biased.

It follows that if significant marker effects remain undetected or unaddressed, conservation and management actions based on those results might not be appropriate. In addition, recent guidelines established by institutional animal care and use committees require that marking protocols minimize pain and stress to study animals.

If researchers collectively ignore the development, evaluation, and application of animal markers acceptable to such committees, and fail to publish results of studies not finding significant effects, then some research might be needlessly jeopardized or precluded. The effects markers can have on animals and to examine critically the treatment of potential marking effects by ecologists. We use the word effect to mean unusual or abnormal behaviour, an abnormal function, or abnormal reproduction or survival.

We use significant to indicate statistical results and important to indicate an observed effect and implication for studies. We emphasize the shortcomings of various marking techniques to animal biology. Our discussion is restricted to effects of markers, and thus does not include a specific review of handling effects.

Furthermore, we do not present results specific to causes of pain or stress because essentially no data exist from wildlife. First, we present the variety of marking techniques that are available for, and explore possible implications of markers on, various taxonomic groups. Next, we review recently published articles to examine how researchers consider potential marking effects. Finally, we discuss how potential marking effects can be minimized and evaluated in future studies.

Consistent with the theme of this volume, the approach we have taken is often critical of existing information and protocols. However, such an approach is necessary if researchers are to improve the overall quality of data being generated from ecological studies. Nietfeld described available marking techniques (excluding marking with radiotransmitters) and generally reviewed marking techniques for vertebrates (excluding fish). Samuel and Fuller provided similar information about radiotransmitters.

Stonehouse edited a book about animal marking, and other overviews dealing with selected vertebrate groups include Stasko and Pincock, Wydowsky and Emery, and Parker et al.for fish; Ferner for amphibians and reptiles; and Marion and Shamis, Calvo and Furness, and Bub and Oelke for birds. These sources will lead the reader to the literature dealing with many species, many marking methods, and various considerations associated with different techniques, different species, and study objectives.

WHICH MARKERS TO USE?

It is worthwhile to reiterate some important factors that Nietfeld et al. and others noted as important when deciding which markers to use for a study. Expense can be an important consideration because marking materials can range widely in cost (e.g., tags versus radiotelemetry via satellites).

The procedures required to initially capture and mark animals and to obtain results from intensive field observations or recapture efforts also are important. Markers should be easily assembled and attached, recognized in the field, and durable enough to remain functional throughout the study.

Additionally, all marking techniques should result in minimum pain or stress to the animal during application and use. Finally, markers should not cause abnormal behaviour or affect survival. Clearly, it is difficult to address all these criteria satisfactorily before the initiation of a study, so some marking has undesirable effects on animals and research results. The adverse effects of marking often are speciesspecific and might occur only in conjunction with certain behaviour (e.g., courtship) or environmental conditions (e.g., extreme temperature). Also, the magnitude and importance of such effects are highly variable among marker types. We present examples of marking techniques and their effects on vertebrate species. This material will help address questions about adverse effects that were raised by Young and Kochert and Nietfeld et al.: Does the information obtained from the study justify marking of animals?

Can the effects of marking be identified during data analysis? If marking effects are accounted for in the analysis, can the study objectives still be achieved? Such questions should be posed at the outset of any study involving the marking of animals. If one or more answers to these questions is negative or unknown, an alternative marker should be sought or the effects of the marker under consideration should be evaluated thoroughly.

EFFECTS OF MARKERS AMONG TAXA

We reviewed a sample of articles that had as a primary objective the evaluation of marker effects. The articles consisted of qualitative or quantitative assessments of the effect of specific marker types on study animals. We acknowledge that marker evaluation studies probably are biased toward those showing effects because results indicating no effects might be published less often.

This implies that our sample of the literature overestimates the occurrence of marker effects in evaluation studies. However, the objective of our review is not to determine how often marker effects occur, but rather to provide examples of the range and diversity of negative effects among marker types, species, and sex, and thus encourage biologists to consider seriously the effects of marking animals. Our review begins with these examples, presented by taxonomic group in the following sections and associated tables.

FISH

Tagging

Marking has been used widely in fish population

estimation; accordingly, the earliest tests evaluating marker effects in vertebrates occurred in fish.

Historically, most evaluations of marking effects were anecdotal, but by the 1940s researchers were suspicious of the potential effects of markers and thus began evaluating their merit in the field. Early fish research often involved the use of commercially made plastic or metal tags, and fish tagging was considered an effective marking system because tags were inexpensive, easily applied and seen, and rarely lost by tagged fish.

However, studies evaluating potential effects of tags often found that tags altered aspects of fish biology. For example, several field studies used mark—recapture techniques and concluded that tags reduced survival and growth of fish.

In some situations, negative effects persisted throughout the life of a fish, whereas in others, the effects seemed to be short-lived. Tagged fish were found to experience reduced swimming ability because of increasing drag, but not all effects of tagging can be attributed directly to the tags themselves.

For instance, choice of tag placement on the fish's body can elicit marker effects, and it is generally considered that tags placed in and around the mouth may interfere with feeding. It is notable that not all tag evaluation studies have shown negative effects of tagging, and with additional study some tags will be shown to be more appropriate than others.

The recent development of passive integrated transponder (PIT) tags has allowed researchers to mark fish and other vertebrates with smaller tags than those used previously. PIT tags are electromagnetically charged microchips implanted either subcutaneously or intraabdominally, and are read remotely via a portable scanner.

Insertion of PIT tags usually is performed using a syringe, thus eliminating the need for extensive invasive surgery. So far, no negative effect of PIT tags has been found in fish, suggesting that this technique can become an important tool for marking fish and other vertebrates. However, one drawback of PIT tagging is that at present tags can be read only when near a scanner.

Mutilation

Marking by mutilation, usually by fin removal or partial removal, is a permanent marking technique often used by fish researchers. However, fin removal often affects fish growth and survival. For example, evaluations using mark—recapture methods show that fin removal causes lower probability of recapture and, by inference, lower survival. Excision of multiple fins generally appears to be more harmful than single-fin excision, and removal of the adipose fin usually is less harmful than removal of other fins.

Removal of dorsal or anal fins can be particularly damaging, partly because under certain conditions such excisions may predispose some species to bacterial or fungal infections or predation. However, as with tagging, not all studies evaluating fin removal have detected significant effects, suggesting that for certain species or age classes, or under specific conditions, this marking technique could be acceptable. Clearly, evaluation of the effects of fin clipping on fish biology requires more attention, particularly under controlled laboratory conditions.

Radiotransmitters

Radiotelemetry has become an important technique in fishery research, allowing biologists to accurately monitor long-term movements and survival of many species that would otherwise be difficult to study. Transmitter sizes and types available for fish are variable, and they have been attached to animals either externally or internally. Laboratory studies have shown that externally mounted transmitters increase drag and reduce or prevent swimming, particularly in high-speed currents.

It has been suggested that fusiform, lotic fishes are more influenced by external mounts than nonfusiform, lentic, or pelagic species. Internal implantation can be achieved by force-feeding stomach transmitters, or by surgery to attach the transmitter either in the peritoneal cavity or intramuscularly. Implants are more commonly used than external transmitters and have the advantages of lying near a fish's centre of gravity, not being lost or entangled in the environment, and not creating drag forces.

However, these advantages can be offset by reduction in swimming performance, increased handling time, and stress associated with surgery, as well as the higher chance of infection following release. Also, implanted transmitters occasionally can be passively expelled from the body, although sometimes without causing mortality or morbidity.

Some species appear more predisposed than others to postoperative complications and transmitter expulsion, meaning that it may be necessary to tailor surgical technique and specific implantation site to the target species. However, in some species, stomach implants seem to have fewer effects than either external mounts or surgically implanted transmitters.

In all telemetry studies, transmitter size is an important consideration, and smaller transmitters are always more desirable than larger ones from the standpoint of effects on the animal. However, the general question regarding the effects of transmitter mass on fish still must be addressed in controlled studies.

REPTILES AND AMPHIBIANS

Tagging

The use of marking in reptile and amphibian research is fairly new, so fewer studies have evaluated marker effects in these taxonomic groups. Many species of reptiles and amphibians have proven difficult to mark because of their epidermal sensitivity, small size, and potential for tissue regeneration. Tagging of reptiles and amphibians has included various types of branding and the use of polymers, pigments, dyes, and radioactive substances.

Many of these markers are of limited utility because they

were not tested adequately for marking effects; such limitations are particularly important for amphibians, given the sensitivity of their skin. A field test of marking by dye injection did not find any effects on larval amphibians, but a controlled laboratory study did identify stunting in dyed tadpoles.

Although these studies used different dyes, the results call into question previous suggestions that some dyes are largely benign and suggest that laboratory studies might be more sensitive to detection of marking effects. Other colour markers, such as fluorescent paint, often are used to monitor amphibians in the field, despite the fact that such paint apparently has not been evaluated for negative effects on behaviour, physiology, or vulnerability to predation.

Most metal or plastic tags used on reptiles and amphibians were originally designed for attachment to fish or birds. Such tags tend to be large and cumbersome, and their effect on study animals remains largely untested despite early suspicions that they affected behaviour and physical condition.

It is promising that both studies evaluating the effect of PIT tags on reptiles failed to detect effects, and with additional study, they might become the standard for tagging many species of reptiles and amphibians. However, in addition to expense and distance requirements for reading, PIT tags have the disadvantage of being lost at a high rate by some free-ranging reptiles.

Mutilation

Until recently, most studies of reptiles and amphibians used mutilation marking to identify individuals. One of the most common forms of mutilation, toe clipping, has been widely used on lizards, frogs, and salamanders because it provides an inexpensive method of identifying individuals. High frequency of natural toe loss in some populations of free-ranging lizards has been used to justify its use as an acceptable marking tool, but the natural occurrence of missing toes does not indicate that toe loss is not traumatic.

Although toe clipping apparently does not affect the sprint performance of some lizards, another study inferred from the low rate of recapture of toe-clipped toads that the marking technique reduced survival. Clarke also noted that recapture rates were inversely related to the number of toes removed, and that toe-clipped toads experienced reduced dexterity when handling large prey.

In other species, regeneration of clipped toes can occur, thus causing problems associated with misidentification of marked animals. However, despite the potential negative effects of toe clipping on reptiles and amphibians, it has remained a widely used form of marking. Clearly, the effect of this technique on reptiles and amphibians requires additional study.

Radiotransmitters

Movements of reptiles and amphibians occasionally have been monitored using a thread-loaded bobbin that unrolls a trail of thread as the animal travels. More often, however, radiotelemetry is used to monitor movements, behaviour, and physiology of reptiles and occasionally amphibians. For amphibians, the main constraint appears to be related to transmitter size, and as a general rule it is recommended that packages not exceed 10 percent of the body mass of the study animal.

Radiotelemetry is problematic with many reptiles and amphibians, and snakes in particular offer challenges because external mounting is not feasible. Considerable effort has been invested in developing an effective method for implanting transmitters in snakes. However, the value of stomach implants is questioned on the grounds that they may affect aspects of snake behaviour.

For instance, stomach-implanted snakes seem to behave similarly to nonimplanted snakes that have recently ingested food, suggesting that activity patterns of implanted snakes are not representative of those of nonimplanted animals. Stomach transmitters also can affect other behaviors or physiological processes, and it might be that such markers simply are not acceptable in snakes because of effects on the animal.

Alternatively, transmitters can be implanted in snakes

either intraperitoneally or subcutaneously, and these modes of attachment generally appear to be effective.

BIRDS

There is more literature about the effects of marking on birds than for other taxa. Therefore, we provide a sample of recent references for numerous avian marking techniques, and refer the reader to recent reviews by Nietfeld et al. and Calvo and Furness for earlier references.

Bands And Collars[SH]

Selecting correct band size and material is a very important step in the marking process because different band materials and configurations can have different effects on birds. For example, aluminum butt-end and lock-on bands can cause more injury and reduce the probability of recovery or recapture of some birds, compared to stainless steel. Some authors recommend use of incoloy (a metal alloy) bands as a substitute for aluminum bands because aluminum bands can cause abrasion to legs of some bird species.

Young birds whose legs are still growing can be the most subject to harmful effects of improperly fitting bands, but one method alleviating such effects is by using plasticine to fill the space between standard size bands and the leg. With this innovation, young birds can be banded without the risk of the large band injuring the bird or slipping off the leg.

Band colour can influence bird behaviour. Burley noted that for captive zebra finches (Poephila guttata), interactions among the sexes and mortality were influenced by legband colour. Among wild zebra finches, Zann found no differences in survival or body condition associated with legband colour, but in one colony females that paired with red-banded males laid more eggs than females paired with males not banded with red.

These results and others reveal that the effects of colour banding can be complex because they vary by species and experimental and environmental conditions. The conspicuousness of colour bands has been enhanced by attaching streamers of the same colour to the band. However, the durability of streamers and fading as well as birds attempting to remove the streamers are problems.

Colour marking also has been accomplished by placing colored tape that contrasts with the plumage on several adjacent flight or tail feathers. Also, feathers can be dyed or painted to enhance detectability, or a portion of colored feather can be used to replace a natural feather. However, like colour banding, colour marking of feathers can cause either significant or negligible effects, depending on various conditions.

Neckbands, which are similar to legbands, have been used on long-necked bird species because they are more easily seen and read. In some cases, neckbands have been found to affect bird survival. Patagial tags, also known as wing tags, are used to enable identification of individual waterfowl.

However, in a study of American coots (Fulica americana), patagial tags were associated with loss of body mass when compared with neckbanded controls. In other species, the use of wing tags may result in wounding, changes in migration times, and reduced reproductive success.

There are other types of bird markers, many having been shown to have effects on birds. For example, titanium dioxide, which was a useful marker on some species, was found to be deleterious to several raptor species. Fluorescent bead markers can be spread in water, from which they attach to waterfowl, and apparently cause no irritation or detectable physical change in the birds. However, the overall efficacy of the marker is subject to exposure time, bird activity, marker transfer among birds, and equipment required for recognition of marked animals.

Radiotransmitters

Radiotelemetry is an important tool for the study of avian biology. Radiomarking and use of recording devices (e.g., for flight velocity, diving depth) on birds has generated considerable study of effects of these markers because they are large compared to most other bird markers. The effect of transmitter or device size on birds can be influenced by where the package is placed or how it is attached. Packages have been placed on legs, necks, wings, backs, retrices and other feathers, under the skin, in the body, by banding, collaring, wing tagging, harnessing, gluing, tying, suturing, clamping, and implanting.

Since the earliest uses of radiomarking in birds, it has been recognized that the transmitter attachment method can affect a variety of aspects of behaviour and survival. For instance, neck collars were shown to be effective in some cases, but in other cases their use was accompanied by negative effects. In one series of studies, tail-mounted transmitters did not affect mass or survival of northern goshawks, but it is understood that such transmitters must be light (2 percent or less of bird body mass), thereby limiting battery size and transmitter longevity.

Reid et al. describe a method for replacing a tailmounted transmitter when individuals can be recaptured. As with legbands and dyes, transmitter colour also must be considered: Wilson and Wilson and Wilson et al. V found that penguins pecked significantly less at black recorders (attached to the dorsal feathers by tape) than at other colors.

Radiotransmitters are similar to other markers in having variable effects that are influenced by attachment method, and the bird's species, age, and sex are necessary considerations. In an effort to securely attach large packages to birds, a number of researchers have experimented with various harness designs to hold transmitters on the bird's back. Several authors failed to show differences in survival or behaviour of birds carrying various harness transmitter packages, but other examples show that such transmitters can have negative effects on bird behaviour or survival.

In an attempt to minimize the deleterious effects of harnesses on birds, researchers have experimented with different types of implants. Partialimplant as well as full-implant transmitters have been used with some success, but such implants can result in short-term preening over the incision site and cause low rates of seroma and infection.

One constraint with transmitter implants is that

transmission distance is reduced when the antenna is implanted in the abdominal cavity, although some researchers have developed a technique to exit the antenna from the body, thereby augmenting transmission distance.

Since the mid-1980s several investigators have considered the effects of radiomarking on bird energetics. This issue is important for birds because radiomarking compounds the increase in energy required to carry additional mass by adding aerodynamic or hydrodynamic drag.

Streamlining of transmitters reduces aerodynamic drag and therefore minimizes their negative effects. Although some researchers failed to detect effects of transmitters weighing less than 4 percent of body mass, others indicate that, either for different species or larger transmitters, radiomarking can affect bird metabolism. This may be particularly important for large birds because they have proportionally less surplus power than smaller birds.

MAMMALS

Tagging

Although studies of mammals often involve marking, marker effects have been evaluated in few instances. This appears to be particularly true for externally mounted metal or plastic tags, with few studies evaluating effects of such markers despite their widespread use. Internal PIT tags have been used in several species of mammals and evaluation tests so far have failed to detect significant negative effects.

Mutilation

Toe clipping is a widely used tool for marking small mammals, and many studies have evaluated the effects of this technique on survival and body condition. Several studies have detected effects of toe clipping, but similar numbers of studies have failed to observe negative effects. In one case, different studies on the same species provided conflicting results, suggesting that study methodology can influence outcome of marker evaluation studies.

In two cases, significant effects of toe clipping were restricted to a particular sex, implying that not all animals are equally vulnerable to negative effects. However, the prevalence of toe clipping as a tool for marking small mammals necessitates that additional studies address its potential effects.

Other forms of mutilation include freeze branding, tattooing, and fur clipping, but in most cases effects of these markers have not been evaluated. However, Stewart and Macdonald did show that European badgers (Meles meles) could be effectively marked via fur clipping and that the mark has no effect on badger body condition. However, the applicability of this technique in colder climates, where loss of guard hairs may affect thermoregulation, requires further study.

Radiotransmitters[SH]

Radiotelemetry of mammals can involve either external or internal attachment of packages. We found that small mammals had received more attention than other groups in marker evaluation studies and that only recently had terrestrial carnivores received consideration for potential transmitter effects.

Most studies we surveyed failed to find significant effects of transmitters, but discrepancies among studies performed on the same species were noted. For instance, the meadow vole (Microtus pennsylvanicus) was subjected to four evaluation tests:

Bertaux et al. did not find a negative effect of transmitters (6.7 to 9.0 percent of body weight) on vole energetics, but other studies showed that transmitters affected vole activity patterns. The fact that differential activity of radiomarked voles was not detected as higher energy expenditure highlights the difficulty associated with attempting to generalize study results.

It also has been shown that effects of radiotransmitters on small mammal behaviour and movements often are either short-term or specific to a particular sex, but the potential demographic implications of such marker effects have not been assessed.

CRITIQUE OF MARKER EVALUATION STUDIES

A primary shortcoming of many marker evaluation studies is experimental design. Sometimes this is manifested as a lack of appropriate controls (i.e., unmarked animals) to which marked animals can be readily compared. The lack of control animals was common during the initial period of marker evaluation studies, but more recent evaluation studies also possess this flaw.

In other cases, authors include in the design a cohort of previously marked or alternatively marked animals as controls. However, this approach assumes that "control" animals are representative of the unmarked population even though alternative markers might cause important effects on their own. In this case, any comparison of marked versus "control" animals could result in an underestimation of effects of the targeted marker.

In some studies, control animals are not subjected to the same handling procedure as marked animals, thereby making marking effects indistinguishable from those of handling. This can be particularly problematic in situations where handling causes significant stress or long-term effects, and researchers may find it difficult to identify which procedure (marking or handling) requires modification.

However, some studies have correctly subjected controls to all the same handling procedures as the marked sample, thus allowing a more rigorous evaluation of the effects of the marker itself.

Marker evaluation studies often have sample sizes that are simply too small to detect a reasonable difference between marked and unmarked samples. Inadequate statistical power increases the likelihood of committing a type II error, thereby increasing the chance of failing to reject a null hypothesis of no significant marking effects when effects actually occur.

Marker effects tend to be more readily detected in the laboratory because field studies often have smaller sample sizes and larger within-sample variance. Many field studies for which marker evaluation is apparently an offshoot, or those that evaluate marker effects on large mammals, lack statistical power. Thus determining the detectable effect size and statistical power associated with a given marker evaluation study should be a necessary precursor to implementation of that study. Also, whenever possible, studies probably should be initiated under controlled laboratory conditions to reduce confounding effects of the environment. However, laboratory studies should be followed by evaluations in the field.

A common characteristic of marker evaluation studies is the use of subjective or qualitative measures of marking effects. Without rigorous statistical treatment of measured effects, results of such studies are of limited utility. Also, indices are sometimes used to infer direct effects (e.g., calculating capture—recapture rates to infer marking effects on survival), but if the index also measures other aspects of species biology (e.g., dispersal), such inferences might be spurious.

Other evaluation studies are too short to derive meaningful conclusions regarding long-term effects, even though the latter effects may very well be the most demographically significant. Finally, sometimes statistically significant results are not considered to be biologically important because they are too small or uncommon. Each of these approaches reduces the likelihood of identifying marker effects that may adversely affect the animal or the study results.

Our review of the marker evaluation literature reveals that a marker can affect a variety of aspects of animal biology, and that different types of evaluations provide different results. When biologists plan an evaluation of marker effects or when they interpret and apply results from previous evaluations, they must make decisions about which methods of evaluation are most appropriate for their objectives and subject species.

Also, they must decide which results are most relevant to assessing the importance of an effect on the animals and their study objectives. For example, a biomechanical analysis of the effect of a marker provides an estimate of how much extra energy is needed to carry the marker, but it does not determine whether that increase in energy expenditure has other biological implications for the animal, such as reduced food delivery to young. A metabolic measurement might not indicate a significant change in O 2 or CO 2 between the marked and unmarked animals. Should we thus conclude there is no effect?

Observations of behaviour of the same animal might reveal that the marked animals spend more time resting than unmarked animals. All these evaluations could produce "significant" results, and yet contradict each other or provide different types of information. It is the responsibility of the researcher to conduct or consider the most appropriate marker evaluations relevant to the study objectives and the well-being of the study animals. The biologist must decide which effects are important.

CURRENT GUIDANCE AVAILABLE FOR CHOOSING MARKERS

Numerous criteria must be considered when selecting markers to be used in a given ecological study, including potential effects of markers on animals. Researchers can review the literature, refer to colleagues, or consult with marker manufacturers or merchants. Limited guidance also is available through guidelines published by scientific journals and government agencies and in general references.

In addition, several professional zoological societies have published guidelines for the use of animals in field research. In general, these societies suggest that markers incur as little pain as possible and not restrict excessively behaviour, physiology, and survival of study animals. We provide a brief overview of some current recommendations provided by zoological societies

TAGGING AND MUTILATION

Tags used on fish, reptiles and amphibians should be of appropriate size and shape, but the use of tags that protrude from the body or are brightly colored is discouraged. It is recommended that for birds all bands be of appropriate size, but the use of nasal disks, saddles, patagial markers, dyes, and ultraviolet markers is discouraged.

Fin clipping is suggested as having minimal impact on survival and social structure of fish, and it is recommended as an appropriate technique if the specific fins to be clipped are expendable by the target species.

Free-ranging reptiles and amphibians should not be toe-clipped unless the technique has been shown not to impair normal activity in the target species or a close relative, whereas for mammals it is recommended that all types of mutilations be avoided. Although birds are occasionally marked via mutilation (i.e., nail clipping, web-punching, feather clipping), no guidelines for use of this technique have been provided.

Marking fish, reptiles, and amphibians using techniques such as tissue removal, branding, freeze branding, and electrocauterization is generally acceptable, but the use of tattoos and paint is less desirable because of problems associated with dye visibility and legibility.

Although fish can be marked with paint, the technique is discouraged for use on amphibian skin, for which nontoxic stains and dyes should be used. In cases where toxicity is unknown, laboratory trials should be undertaken before any field use. Few guidelines are provided for the use of brands, dyes, or paints with birds and mammals.

Radiotransmitters

Most professional zoological societies address the issue of radiotransmitters specifically. Many fish, reptile, and amphibian species are not suitable for radiotelemetry because of their small size. However, for species that are amenable to telemetry, stomach implants and internally mounted transmitters should be small and coated with a biologically inert coating, and not interfere with physiology and behaviour.

Externally mounted transmitters should be shaped and attached to reduce chances of entanglement, irritation, or constriction. In the case of large birds, it is suggested that radiotransmitters weigh less than 1 percent of body mass to reduce negative effects on biomechanical performance. For smaller birds transmitters should not exceed 5 or 10 percent of body mass. Before use in the field, biologists should observe individuals in captivity to evaluate effects of radiomarking on behaviour. For reptiles, amphibians, and most mammals, it is recommended that transmitters not exceed 10 percent of body mass.

CRITIQUE OF GUIDELINES AVAILABLE FOR CHOOSING MARKERS

In general, the guidelines provided by zoological societies are too general for choosing a specific marker for a given study objective or species. Some recommendations made by professional societies even appear to ignore the findings of previous marker evaluation studies. For instance, fin removal is recommended by several fish societies as an appropriate method of marking many species, despite numerous instances in which the technique has been shown to affect fish biology.

Also, it is recommended that mass of transmitters never exceed 10 percent of body mass of vertebrates, even though some transmitters weighing less than this have been shown to produce negative effects. Given that, at least for birds, the effect on flight power increases with body mass, the 10 percent threshold is clearly an arbitrary construct that does not apply to all species. Furthermore, a given mass or drag has different effects depending on the type of flight. Therefore, general guidelines (e.g., 10 percent, 5 percent, or 3 percent) can be misleading.

Although for terrestrial animals costs associated with carrying a load increase linearly with load mass, intra- and interspecific differences in biology should preclude the establishment of a threshold that transcends taxonomic groups. Societies should assume a more comprehensive responsibility to indicate to their members possible shortcomings associated with various marking procedures, although it remains the responsibility of individual researchers to address the potential effects ot markers on their study species.

Where information is lacking on a particular marker or its effect, zoological societies should encourage additional evaluation studies. It is imperative that markers used in research be acceptable to animal care and use committees before project initiation, and zoological societies should strive to harmonize their standards.

With those set by such committees and disseminate information regarding acceptable markers and protocols to committees. Otherwise, research using newly tested markers might be needlessly precluded because of an uncertainty regarding marker effects.

RECENT ECOLOGICAL STUDIES

In light of the measurable and obvious effects of markers to numerous vertebrate species, we examined the treatment of potential marking effects in a sample of recent peer-reviewed literature. Specifically, we assessed the frequency with which researchers addressed potential marker effects, either by using methodologies that reduced potential effects or by testing for such effects qualitatively or statistically.

We surveyed nine journals that publish studies on a broad range of taxonomic groups for articles in which vertebrates had been marked; our survey included only articles in which the primary objective was to address general issues related to animal biology rather than to study marking effects.

We found that in most instances (90 percent, n = 238), authors did not address potential effects of marking, or at least did not report any such consideration in the article. Undoubtedly, some authors attempted to minimize or evaluate marker effects but chose not to report these efforts in the article; however, we suspect that such cases were uncommon.

We found no indication that the failure to evaluate marker effects was weighted toward a given taxonomic group or journal. In 3 percent of cases authors assumed explicitly that markers did not affect animals. Such assumptions were usually based on previous reports showing no significant marking effects in the target (or related) species or on requirements of the statistical method chosen to analyse the data.

Only 7 percent of the articles surveyed included information directly pertaining to potential marking effects. Some authors attempted to minimize marker effects by allowing postmarking recovery to take place in captivity.

Although this approach might be effective when handling and marking are stressful or invasive (i.e., internally mounted radiotransmitters), it might also bias study results if captivity affects behaviour or survival after release. For species that are highly vulnerable to stress from captivity, prolonged recovery under controlled situations could be less desirable than immediate release.

Other attempts to minimize marker effects included adjusting markers to fit individual animals or using expandable and breakaway markers. Some authors justified the use of a marker by describing qualitatively how animals appeared to behave normally after marking.

Subjective evaluation of potential marking effects is a good first step, but whenever possible quantitative measurements of the biology of marked versus unmarked animals or comparisons of preand postmarking behaviour of individuals should be included.

When markers were suspected of exerting short-term effects on animals, researchers excluded data obtained during an arbitrary period. However, this practice can result in data bias if marking causes subsequent increases in mortality of young or frail animals; the result would be an inflation of survival estimates. Two studies evaluated the effect of one marker type using an alternatively marked cohort as control, without validating that negative effects of the alternative marker did not occur.

We found no indication from our survey that authors addressed potential long-term effects of markers, although one study inferred from the lack of external damage in study animals that long-term tagging mortality was absent or negligible. In another case marked and unmarked animals were monitored and data from both cohorts were apparently pooled, but no mention was made of trying to evaluate possible marker effects.

It is also notable that almost all (more than 90 percent, n = 16) instances in which authors addressed marking effects the studies involved animals that were marked with radiotransmitters. This implies that various forms of mutilation and tagging or banding are either assumed by researchers to have little or no effect or else not afforded the same scrutiny in the editorial process.

To summarize, our review of ecological studies found that in general authors tended to overlook or not acknowledge potential marking effects. In few cases potential marking effects were addressed, but usually only in a superficial manner or one that could produce bias in the resulting data.

This review clearly indicates that there are a large number of reported marking effects from commonly used markers, there is a lack of thorough guidance from zoological societies in choosing a marker, and there is a common failure to address potential marking effects in the ecological literature. Thus, we provide the following recommendations for using markers and for additional research into the effects of marking wildlife.

PROTOCOLS AND TECHNOLOGICAL ADVANCES

Depending on study objectives, target species, and possible problems associated with the handling and marking process, it is not always necessary or desirable to apply marks to animals. For example, in some studies genetic or mineral markers in animal tissues, excreta, or blood, may allow for identification of individuals or populations. However, reliable identification using genetics is, at present, labour intensive and costly, can require recapture, and is largely untested under field conditions.

As another alternative, it might be possible to use naturally occurring variable colour markings or unique morphological features for recognition of individuals. For example, some amphibian species are amenable to visual recognition of individuals, and photographs or sketches can be used to record characteristics of individuals in a population. Individual recognition via natural markings also has been used occasionally for fish and some bird species, but has probably found its greatest utility with large mammals.

For mammals, researchers usually rely on unique facial features to identify individuals; in most cases the method appears reliable although validation is desirable and there are limits to the number of animals in a population that are individually recognizable. Pennycuick reviewed limitations and difficulties associated with the use of natural markings in free-ranging animals; where the use of this technique meets study objectives, it is more desirable than artificial marking.

It is essential that field personnel be adequately trained in the proper handling and marking of animals. This may involve practice with captive individuals of the targeted or surrogate species, or animal models.

Improper marker application by poorly trained field personnel can result in serious problems for study animals, and well-trained personnel not only minimize the potential for negative marker effects but also are able to process and release study animals more rapidly, thereby reducing potential negative effects of handling and marking.

Although significant progress has been made in the development of less harmful marks, additional steps are necessary. Few technological developments can contribute to minimizing effects of mutilations, although routine use of surgical instruments and sterile conditions lessens occurrence of disease and mortality.

Development of smaller tags and bands can contribute to less invasive marking. In the past, one constraint of reducing tag size was the ability to read numbers or colour markings identifying individual animals. However, carefully designed protocols can increase.

The use of PIT tags eliminates size constraints for many animals. However, other tags also hold promise, such as passively applied (i.e., usually ingested) group recognition markers. These may prove particularly useful when animal handling is undesirable and individual recognition is not necessary.

Radiotransmitters probably will remain an important and

widely used form of marking, and technological developments should include aerodynamically or hydrodynamically shaped, smaller, and lighter packages and development of new attachment methods.

Because most of the bulk caused by transmitters is attributable to batteries, packages could be significantly lightened if smaller high-capacity batteries were developed. However, in some cases significant reductions in transmitter size or mass are not possible without forgoing battery longevity.

Thorough testing and reporting of animal marking procedures provides other investigators information useful for choosing among various attachments and trying them with a new species or under new circumstances. Detailed descriptions of procedures and modifications to procedures provide potential users with additional options.

Finally, the results of applying various attachments in field situations are very important and should include information about effects of the capture and handling that must accompany marking. Sometimes the procedures associated with capture and handling can be as important a component in the marking process as the marker itself.

Our review reveals such a variety of marker effects (depending on marker type, species, age, sex, attachment method, environment, and season) that careful development, testing, documentation, and reporting of experience with markers will continue to be very useful in identifying and dealing with marker effects. Zoological societies should assume the responsibility for setting reasonable standards for the marking of animals by providing detailed guidelines.

MARKER EVALUATION STUDIES

We recommend increased efforts to test for effects of markers, changes in existing protocols and recommendations, and more publication of the results of marker evaluation studies. Evaluation studies are a critical step in the marker validation process, but many commonly used markers have been evaluated for only a few species. Our review highlighted several specific deficiencies in our understanding of marker effects, such as effects of radiotransmitters in fish, efficacy of internally versus externally mounted transmitters (particularly stomach implants), influence of external tags on amphibians and mammals, and effects of transmitters on large mammals, particularly carnivores.

It will not be possible to evaluate the effects of all markers in all species, but results of evaluation studies will be more useful if researchers focus their initial evaluation studies on broadly applicable markers and species that share several biological features with other species of interest.

We think that PIT tags will become very important for marking animals in the future, and thus urge researchers to undertake the necessary evaluation studies without delay. All evaluation studies should be designed to measure marker effects across ages and sexes.

Evaluation studies must include comparison to unmarked (or alternatively marked) controls and adequate sample sizes. Statistical power analysis can be used to calculate the minimum sample size required to detect significant effects. Captive or laboratory conditions should be used whenever possible, although researchers should ensure that biological parameters being measured are not significantly affected by captivity.

Ultimately, results of captive studies should be validated in the field. Also, evaluation studies should take advantage of sensitive techniques for estimating subtle negative effects, such as biomechanical modeling and doubly labeled water. Marker evaluation studies should attempt to address indirect effects of markers on survival (e.g., predation, starvation, disease) and behaviour (e.g., altered foraging or parental abandonment of juveniles to compensate for higher energy demands), and focus not only on short-term but also on long-term effects of markers.

Marker evaluation studies should be published in the peer-reviewed literature. It was not possible for us to determine whether a publication bias existed in our sample of marker evaluation studies, but we suspect a tendency exists for publishing studies that show significant negative effects over those failing to show effects. Such a bias can misrepresent the effect of markers, and it inhibits dissemination of information about the best marking methods.

Clearly, some journals must encourage authors to perform and publish marker evaluation studies. A similar suggestion was proposed to overcome the apparent publication bias in biological studies showing nonsignificant results. Publication of all marker evaluation studies would provide researchers and animal care and use committees with current knowledge regarding acceptable animal markers.

It remains the responsibility of reviewers and editors to ensure that appropriate marker evaluation studies are cited or conducted in the published ecological and wildlife literature. In some cases, it may be sufficient to justify the use of a given marker by citing previous studies that have failed to detect effects, but when researchers are applying a marker to a new (or unrelated) species, or using a marker that has not been evaluated previously, appropriate tests must be undertaken.

Again, experimental design and sample size are paramount considerations for tests. Publication of technique validation studies is common in other scientific fields and should be encouraged in animal ecology.

Because current funding sources often do not support technique evaluation studies, researchers must find other sources of funding for such work, such as private animal welfare organizations, and they must encourage traditional sources of funding to support evaluation of marking methods. Only through additional studies and published results will it be possible for colleagues to thoroughly evaluate potential marker effects and choose the most appropriate markers.

Chapter 2

Animal Home Ranges

DEFINITION OF HOME RANGE

Most animals are not nomadic but live in fairly confined areas where they enact their day-to-day activities. Such areas are called home ranges. Burt (1943:351) provided the verbal definition of a mammal's home range that is the foundation of the general concept used today: "that area traversed by the individual in its normal activities of food gathering, mating, and caring for young.

Occasional sallies outside the area, perhaps exploratory in nature, should not be considered part of the home range." This definition is clear conceptually, but it is vague on points that are important to quantifying animals' home ranges. Burt gave no guidance concerning how to quantify occasional sallies or how to define the area from which the sallies are made.

The vague wording implicitly and correctly allows a home range to include areas used in diverse ways for diverse behaviors. Members of two different species may use their home ranges very differently with very different behaviors, but for both the home ranges are recognizable as home ranges, not something different for each species.

How does an animal view its home range? Obviously, with our present knowledge we cannot know, but to be able to know would provide tremendous insight into animals' lives. Aldo Leopold (1949:78) wrote, "The wild things that live on my farm are reluctant to tell me, in so many words, how much of my township is included within their daily or nightly beats.

The ratio between the size of their universe and mine,

and it conveniently begs the much more important question, who is the more thoroughly acquainted with the world in which he lives?" Leopold continued, "Like people, my animals frequently disclose by their actions what they decline to divulge in words."

We do know that members of some species, probably many species, have cognitive maps of where they live or concepts of where different resources and features are located within their home ranges and of how to travel between them. Such cognitive maps may be sensitive to where an animal finds itself within its home range or to its nutritional state; for example, resources that the animal perceives to be close at hand or resources far away that balance the diet may be more valuable than others.

From extensive research on optimal foraging, we know that animals often rank resources in some manner. Consequently, we might envision an animal's cognitive map of its home range as an integration of contour maps, one (or more) for food resources, one for escape cover, one for travel routes, one for known home ranges of members of the other sex, and so forth.

Why do animals have home ranges? Stamps (1995:41) argued that animals have home ranges because individuals learn "site-specific serial motor programs," which might be envisioned as near reflex movements that take an animal along well traveled routes to safety.

These movements should enhance the animal's ability to maneuver through its environment and thereby to avoid or escape predators. Stamps argued that the willingness of an animal to incur costs to remain in a familiar area implies that being familiar with that area provides a fitness benefit greater than the costs.

For animals with small home ranges that live their lives as potential prey, Stamps's hypothesis makes sense. However, many animals, especially predatory mammals and birds, have home ranges too large and use specific places too seldom for site-specific serial motor programs to have an important benefit.

Site-specific serial motor patterns of greatest use to a predator would have to match the escape routes of each prey individual, but each of these might be used only once after it is learned. The reason that animals maintain home ranges must be broader than Stamps's hypothesis.

Nonetheless, Stamps has undoubtedly identified the key reason that animals establish and maintain home ranges: The benefits of maintaining a home range exceed the costs. Let C D be the daily costs for an animal, excluding the costs, C R, of monitoring, maintaining, defending, developing, and remembering the critical resources on which it based its decision to establish a home range. In the long term, C D plus C R must be equal to or less than the benefits, B, gained from the home range,

$$C_D + C_R \leq B$$

Costs and benefits must ultimately be calculated in terms of an animal's fitness, but if the critical resources are food, then costs and benefits might be indexed by energy. If the benefits are nest sites or escape routes, energy is not an adequate index. If C D plus C R exceeds B for an animal in the short term, then the animal might be able to live on a negative balance until conditions change.

If C D plus C R exceeds B in the long term, then the animal must reduce C D, or C R, both of which have lower limits. C D generally cannot be reduced below basic maintenance costs, or basal metabolism; however, hibernation and estivation are methods some animals can use to reduce C D below basal metabolism.

Reducing C R might reduce B because benefits can be experienced only through attending to critical, local resources, which is C R. If C R can be reduced through increased efficiency, B need not be reduced when C R is reduced or need not be reduced as much as C R is reduced. Ultimately, in the long term, if C D + C R > B then the animal cannot survive using local resources. If the animal cannot survive using local resources, it must go to another locale where benefits exceed costs, or it must be nomadic and not exhibit site fidelity.

Because maintaining a home range requires site fidelity, site fidelity can be used as an indicator of whether an animal has established a home range. Operational definitions of home ranges exist using statistical definitions of site fidelity.

The goals of such definitions are good but the methods sometimes fail to define home ranges for animals that exhibit true and localized site fidelity. For example, Swihart and Slade used data for a female black bear (Ursus americanus) and determined that she did not have a home range because the sequence of her locations did not show site fidelity as defined by their statistical model.

However, the bear's locations were strictly confined for 3 years to a distinct, well-defined area. Consequently, researchers must sometimes use subjective measures of site fidelity, to augment objective measures that sometimes fail, probably because statistical models have assumptions that are not appropriate for animal movements. Nonetheless, tests of site fidelity should be disregarded only when other objective approaches to site fidelity exist.

An animal's cognitive map must change as the animal learns new things about its environment and, hence, the map changes with time. As new resources develop or are discovered and as old ones disappear, appropriate changes must be made on the map. Such changes may occur quickly because an animal has an instantaneous concept of its cognitive map.

A researcher, in contrast, can learn of the changed cognitive map only by studying the changes in the locations that the animal visits over time. An animal's home range usually cannot be quantified, practically, as an instantaneous concept because the home range can only be deduced from locations of an animal within its home range and the locations occur sequentially.

Thus, for most approaches, a home range must be defined for a specific time interval (e.g., a season, a year, or possibly a lifetime). The longer the interval, the more data can be used to quantify the home range, but the more likely that the animal has changed its cognitive map since the first data were collected.

In addition, no standard exists as to whether one should include in an animal's home range areas that the animal seldom visits or never visits after initial exploration. Many researchers define home ranges operationally to include only areas of use. Nonetheless, animals may be familiar with areas that they do not use.

An arctic fox (Alopex lagopus) may be familiar with areas larger than 100 km2, yet use only a small portion (ca. 25 km2) where food is concentrated. Areas with no food are not visited often, if ever, despite and because of the animal's familiarity with them. Should such areas be included in the fox's home range? Other areas with food might not have been visited in a given year simply by chance.

Should those areas be included in the fox's home range? Pulliainen asserted that any area larger than 4 ha (an arbitrary size) not traversed by the Eurasian martens (Martes martes) he and his coworkers followed should not be included in the martens' home ranges. Through a winter, a marten crosses and recrosses old travel routes, leaving progressively smaller and smaller areas of irregular shape surrounded by tracks.

Pulliainen presumed that a marten's radius of familiarity, or radius of perception, would cover an area of 4 ha or less. But how wide might an animal's radius of perception be? Some mammals can smell over a kilometre, see a few hundred meters, but feel only what touches them.

Which radius should be used, or should a multiscale radius be used? In addition, areas not traversed may have been avoided by choice. Hence, should no radius of familiarity be considered? If we do not allow some radius of familiarity, or perception, around an animal, we are reduced, reductio absurdum, to counting as an animal's home range only the places where it actually placed its feet. Clearly, this is not satisfactory.

Related to this final problem is how to define the edges of an animal's home range. For many animals, the edges are areas an animal uses little but knows; the animal may actually care little about the precision of the boundaries of its home range because it spends the vast majority of its time elsewhere.

Except for some territorial animals, the interior of an animal's home range is often more important both to the animal and to understanding how the animal lives and why the animal lives in that place.

Gautestad and Mysterud and others have noted that the boundaries of home ranges are diffuse and general, making the area of a home range difficult to measure. That the boundary and area of a home range are difficult to measure does not reduce in any way the importance of the home range to the animal and to our understanding of the animal, however.

Even crudely estimated areas for home ranges have led to insights into animal behaviour and ecology, suggesting that home range areas should be quantified. However, we must keep in mind that home range boundaries and areas are imprecise, at least in part, because the boundaries are probably imprecise to the animals themselves.

Territories

A territory is an area within an animal's home range over which the animal has exclusive use, or perhaps priority use. A territory may be the animal's entire home range or it may be only part of the animal's home range (its core, for example). Territories may be defended with tooth and claw (or beaks, talons, or mandibles) but generally are defended through scent marking, calls, or displays, which are safer, more economical, and evolutionarily stable.

Members of many species, such as red squirrels, defend individual territories against all conspecifics, but tremendous variation in territorial behaviour exists. In some species, individuals defend territories only against members of the same sex. In other species, mated pairs defend territories.

In still other species, extended family groups, sometimes containing none—family members, defend territories. Whether territories are defended by an individual, mated pair, or family appears to depend on the productivity, predictability, and fine-grained versus coarse-grained patchiness of the limiting resources.

Members of many species in the Carnivora exhibit

intrasexual territoriality and maintain territories only with regard to members of their own sex. These species exhibit large sexual dimorphism in body size and males of these species are polygynous (and females undoubtedly selectively polyandrous). Females raise young without help from males and the large body sizes of males may be considered a cost of reproduction.

For species that affect food supplies mostly through resource depression (i.e., have rapidly renewing food resources such as ripening berries and nuts or prey on animals that become wary when they perceive a predator and later relax), intrasexual territoriality appears to have a minor cost compared to intersexual territoriality because the limiting resource renews. This cost may be imposed on females by males.

Males of many songbird species defend territories. In migratory species, the males usually establish their territories on the breeding range before the females arrive and a male will continue to defend his territory if his mate is lost early in the breeding season. For these territories, the limiting resource may be a complex mix of the food and other resources that females need for successful reproduction and the females themselves.

In red-cockaded woodpeckers (Picoides borealis) and scrub jays (Aphelocoma coerulescens), however, extended family groups defend territories. Male offspring, or occasionally female offspring, remain in their parents' (fathers') territories. Wolves (Canis lupus), beavers (Castor canadensis), and dwarf mongooses (Helogale parvula) also defend territories as extended families.

Although territorial behaviour might intuitively appear to help clarify the problem of identifying home range boundaries, this is not always the case. The territorial behaviour of wolves actually highlights the imprecise nature of the boundaries of their territories. Peters and Mech documented that territorial wolves scent marked at high rates in response to the scent marks of neighboring wolf packs.

In addition, the alpha male of a pack often ventured up

to a couple hundred meters into a neighboring pack's territory to leave a scent mark. Such behaviour changes a territory boundary into a space a few hundred meters wide, not a distinct, linear boundary. Hence, distinct boundaries of territories are little easier to identify than are boundaries of undefended home ranges.

Animals are territorial only when they have a limiting resource, that is, a critical resource that is in short supply and limits population growth. The ultimate regulator of a population of territorial animals is the limiting resource that stimulates territorial behaviour. Although population regulation through territoriality has received extensive theoretical attention, the general conclusion of such theory is that territoriality can regulate populations only proximally.

The most common limiting resource is food and, for territorial individuals, territory size tends to vary inversely with food availability. For red-cockaded woodpeckers, however, the limiting resource is nest holes. For coral reef fish, the limiting resource is usually space. For pine voles (Microtus pinetorum), the limiting resource appears to be tunnel systems. And for beavers, the limiting resource may be dams and lodges. Wolff warned that the limiting resource may not be food even if it appears superficially to be food.

Territorial behaviour is not a species characteristic. In some species, individuals defend territories in certain parts of the species' range but not in other parts. This is the case for black bears. Similarly, many nectarivorous birds defend territories only when nectar production is at certain levels. To understand why members of these species display flexibility in their territorial behaviour, one must start with the concept that a territory must be economically defensible.

Carpenter and MacMillen showed theoretically that an animal should be territorial only when the productivity of its food (or whatever its limiting resource is) is between certain limits. When productivity is low, the costs of defending a territory are not returned through exclusive access to the limiting resource. When productivity is high, requirements can be met without exclusive access.

The model developed by Carpenter and MacMillen is broadly applicable because it expresses clearly the limiting conditions required for territoriality to exist and it incorporates limits on territory size from habitat heterogeneity, or patchiness. Some approaches to modeling territorial behaviour, such as Ebersole's, Hixon's, and Kodric-Brown's and Brown's, do not express limiting conditions for territorial behaviour but tacitly assume, a priori, that territoriality is economical.

Understanding the limiting conditions for territorial behaviour is important to understanding spacing behaviour and home range variation in many species. Using economic models is a good approach to understanding limiting conditions for territoriality as long as the limiting resources do not change as conditions change.

Otherwise, the limiting resources must all be known clearly for the different conditions under which each is limiting. For example, if a small increase in the abundance of food leads to another resource becoming the limiting resource, that new limiting resource must be understood as well as the importance of food is understood. Researchers must also understand how an economic modeling approach fits into a broader picture, such as how animals use information from the environment to make decisions and how they perceive information.

When productivity of the limiting resource for an individual is very low and close to the lower limit for territoriality, the individual must maintain a territory of the maximum size possible. Such an individual should be completely territorial and not share any part of its territory.

As productivity of the limiting resource for an individual approaches the upper limit for territoriality, however, its territorial behaviour should change in one of two predictable ways. If necessary resources are evenly distributed in defended habitat, then the individual should maintain a smaller territory than in less productive habitat.

If the individual's resources are distributed patchily and balanced resources cannot be found in a small territory, then it might exhibit incomplete territoriality. The individual might maintain exclusive access only to the parts of its home range with the most important resources. Coyotes, European red squirrels, and perhaps red-cockaded woodpeckers exhibit just such a pattern of partial territoriality and defend only home range cores in some habitats. Alternatively, an individual might allow territory overlap with a member of the opposite sex.

Food appears to be the limiting resource that stimulates territorial behaviour by many animals and territorial defence decreases in those individuals as productivity or availability of food increases. Much research has been done on nectarivorous birds, voles, and mammalian carnivores. Black bears and nectarivorous birds switch quickly between territorial and nonterritorial behaviour when productivity of food moves across the lower or upper limits for territoriality, respectively.

Territorial behaviour by members of several species can be predicted from variation in the productivity of food, which is good evidence that food is the limiting resource that stimulates territorial behaviour for those animals. For European badgers (Meles meles), territory configuration can be predicted from positions of dens without reference to food, indicating that the limiting resource is den sites.

However, no studies have rejected all other possible limiting resources. Wolff argued strongly that only offspring are important enough, and can be defended well enough, to be the resource stimulating territorial behaviour. For the black bears, adult females with and without young and adult and juvenile bears all responded in the same manner to changes in food productivity and also responded in the same manner to home range overlap with other female bears.

Were Wolff correct, adult female bears with young would exhibit significantly different responses to food and to other females than do nonreproductive females. In addition, adult female black bears would be territorial in North Carolina, as they are in Minnesota.

For the nectarivorous birds studied by Hixon, birds

defended territories in the fall after reproduction but before and during migration. Were Wolff correct, hummingbirds would not defend territories after reproduction has ceased for the year. If bears, hummingbirds, and other animals use food as an index for the potential to produce offspring, then food can legitimately be considered to be at least a proximately limiting resource.

Fitness is the ultimate currency in biology, and fitness may be affected by one or more limiting resources that need not be offspring or other direct components of reproduction. Evolution via natural selection requires heritable variation that affects reproductive output among individuals in a population. The effects can be via offspring, or they can be via food, nest sites, tunnel systems, or other potentially limiting resources.

ESTIMATING ANIMALS' HOME RANGES

Added to conceptual problems of understanding an animal's home range are problems in estimating and quantifying that home range. We may never be able to find completely objective statistical methods that use location data to yield biologically significant information about animals' home ranges. Nonetheless, our goal must be to develop methods that are as objective and repeatable as possible while being biologically appropriate. When analyzing data, we must use a home range estimator that is appropriate for the hypotheses being tested and appropriate for the data.

Reasons for estimating animals' home ranges are as diverse as research and management questions. Knowing animals' home ranges provides significant insight into mating patterns and reproduction, social organization and interactions, foraging and food choices, limiting resources, important components of habitat, and more. A home range estimator should delimit where an animal can be found with some level of predictability, and it should quantify the animal's probability of being in different places or the importance of different places to the animal.

Quantifying an animal's home range is an act of using data about the animal's use of space to deduce or to gain insight into the animal's cognitive map of its home. These data are usually in the form of observations, trapping or telemetry locations, or tracks. Because at present we have no way of learning directly how an animal perceives its cognitive map of its home, we do not have a perfect method for quantifying home ranges. Even if we could understand an animal's cognitive map, we would undoubtedly find it difficult to quantify. Many methods for quantifying home ranges provide little more than crude outlines of where an animal has been located. For some research questions, no more information is needed. For questions that relate to understanding why an animal has chosen to live where it has, estimators are needed that provide more complex pictures.

An animal's cognitive map will have incorporated into it the importance to the animal of different areas. The most commonly used index of that importance is the amount of time the animal spends in the different areas in its home range. For some animals, however, small areas within their home ranges may be critically important but not used for long periods of time, such as water holes.

No standard approach exists to weight use of space by a researcher's understanding of importance. Therefore, to estimate importance to an animal of different areas of its home range, using any home range estimator currently available, one must assume that importance is positively associated with length or frequency of use, which are measures of time.

MINIMUM CONVEX POLYGON

The oldest and mostly commonly used method of estimating an animal's home range is to draw the smallest convex polygon possible that encompasses all known or estimated locations for the animal. This minimum convex polygon is conceptually simple, easy to draw, and not constrained by assuming that animal movements or home ranges must fit some underlying statistical distribution. However, problems with the method are myriad.

Minimum convex polygons provide only crude outlines of animals' home ranges, are highly sensitive to extreme data points, ignore all information provided by interior data points, can incorporate large areas that are never used, and approach asymptotic values of home range area and outline only with large sample sizes. Because all information about use of a home range within its borders is ignored using a minimum convex polygon, most analyses using this method implicitly assume that animals use their home ranges evenly (use all parts with equal intensity), which is clearly not the case.

One can calculate a minimum convex polygon using the 95 percent of the data points that form the smallest polygon, but this does not avoid the flaws inherent in the method other than the problem with extreme data points.

To construct a minimum convex polygon, a researcher discards 90 percent of the data he or she worked so hard to collect and keeps only the extreme data points. This method, more than any other, emphasizes only the unstable, boundary properties of a home range and ignores the internal structures of home ranges and central tendencies, which are more stable and are important for most critical questions about animals.

CIRCLE AND ELLIPSE APPROACHES

Hayne suggested that to estimate an animal's home range from point location data one should use a circle; Jennrich and Turner and Dunn and Gipson generalized the circle to an ellipse. Circle and ellipse approaches assume that animals use space in a fashion conforming to an underlying bivariate normal distribution. Using a circle to represent an animal's home range assumes that each animal has a single centre of activity that is the very centre, or the two-dimensional arithmetic mean, of all locations. Using an ellipse assumes that each animal has two such centers of activity that are the foci of the ellipse.

An ellipse can be drawn around the two centers of activity for an animal such that it contains 95 percent of the location data. This 95 percent ellipse can also be used as an estimate of the animal's home range.

FOURIER SERIES

In statistics, Fourier series are often used to smooth data, so Anderson developed a home range estimator based on the bivariate Fourier series. Each animal location estimate is treated as a spike in the third dimension above an x–y plane. The Fourier transform estimator smooths the spikes into a surface that estimates an animal's utility distribution.

Both of these estimators accurately show multiple centers of activity that may be considerably removed from the arithmetic mean of the x and y data, but both behave poorly near the edges of home ranges, probably because the location data do not meet assumptions needed to make the transformations. To address the problem of poor estimates of home range peripheries,

Anderson recommended using animals' 50 percent home ranges (the smallest area encompassing a 50 percent probability of use) rather than 95 percent home ranges. Fifty percent is no less arbitrary than 95 percent, but it departs completely from the basic concept of a home range or stretches that concept to its limit by assuming that an animal is on an "occasional sally" 50 percent of the time.

HARMONIC MEAN DISTRIBUTION

Human population densities fall in an inverse harmonic mean fashion from centers of urban areas through rural areas. Consequently, Dixon and Chapman proposed using a harmonic mean distribution to describe animal home ranges. Contours for a utility distribution are developed from the harmonic mean distance from each animal location to each point on a superimposed grid.

The harmonic mean estimator may accurately show multiple centers of activity, but each estimated utility distribution is unique to the position and spacing of the underlying grid. Spencer and Barrett modified the method to reduce the problem of grid placement but a large problem with grid size remains. When a very fine grid is used, the resulting utility distribution becomes a series of sharp peaks at each animal location.

When a coarse grid is used, the utility distribution lacks local detail and is overly smoothed. For many data sets, the harmonic mean estimator actually appears both to exaggerate peaks at animal locations and to oversmooth elsewhere. In addition, the estimator calculates values for all grid points, provides no outline for a home range, and does not provide a utility distribution.

Most researchers choose for the home range outline the contour equal to the largest harmonic mean distance from an animal location to all other animal locations and from this a utility distribution can be calculated. Although this is an objective criterion, it is affected by sample size. Finally, for animal home ranges that have geographic constraints that confine shapes, much area not actually in an animal's home range will be included in the harmonic mean estimate.

Boulanger and White used Monte Carlo simulations and tested the performance of the harmonic mean estimator against the other estimators just discussed. Despite its problems, the harmonic mean estimator was the best of the lot. Luckily, better estimators have since been developed. One set of home range estimators, kernel estimators, appears best suited for estimating animals' utility distributions, and hence home ranges. Another set, fractal estimators, may have promise.

FRACTAL ESTIMATORS

Bascompte and Vilà, Gautestad and Mysterud, and Loehle modeled animal movements as multiscale random walks and analyzed the patterns of locations as fractals. Bascompte and Vilà explained that D, the fractal dimension, can be estimated as

$$D = \frac{\log(n)}{\log(n) + \log(d/L)}$$

where n is the number of steps along a trace of an animal's movements (1 less than the number of locations), L is the sum of the lengths of all steps (total length of the movement), and d is the planar diameter, which can be estimated as the greatest distance between two locations.

For a movement that is a straight line, d = L, so D = 1; a line has one dimension. For a random walk, D = 2; a random walk spreads over a plane and has two dimensions.

For the animals studied by Bascompte and Vilà and Gautestad and Mysterud, the fractal dimensions, D, for movements averaged less than 2. Finding D < 2 means that as they scrutinized their animal location data on smaller and smaller scales, they found clumps of locations within clumps within clumps ad infinitum. The movements of the animals did not spread randomly across the landscape.

Gautestad and Mysterud argued, therefore, that animals use their home ranges in a multiscale manner, which makes ultimate sense. Optimality modeling (giving up time) and empirical data show that animals who forage in patchy environments are predicted to and, indeed, do change their movements dependent on both finescale and large-scale characteristics of food availability.

Thus an animal's decision to remain in or to leave a food patch depends not just on the availability of food within the patch but also on the availability of food across its home range and on the locations of the other patches of food.

In addition, Gautestad and Mysterud showed that if animals move in a manner described by a multiscale random walk that incorporates the multiscale, fractal nature of animal movements, then the estimated home range area should increase infinitely in proportion with the square root of the number of location estimates used to estimate the area of the home range using a minimum convex polygon.

Indeed, the home ranges of several species, quantified using minimum convex polygons, do appear to increase in area as predicted. The predicted relationship between home range area (A MCP, for minimum convex polygons) and the number of locations (n) is

$$A_{MCP} = C.Q(n).n^{1/2}$$

where C is the constant of proportionality, or the scaling factor, and Q(n) is a function that adjusts the relationship for underestimates of A MCP because of small sample size. Curve fitting indicates that

$$Q(n) = \exp\left(6/n^{0.7}\right)$$

for n e" 5. When not calculating home range area from minimum convex polygons, Q(n) should not be used.

Gautestad and Mysterud (1993) interpret C to be a measure of how an animal perceives the grain of its environment. When a grid is superimposed over a plot of an animal's locations, C can be calculated for each cell and 1/C is a descriptor of the intensity of use for each cell.

1/C can be calculated in two ways. Superimpose a grid on a map of a study area such that no cells have fewer than five locations for a target animal (cells with fewer than five locations might alternatively be ignored). Calculate the area of the minimum convex polygon formed by all locations within each cell and use that for A MCP in equation 3.1. Calculate 1/C as

$$1/C = \left[Q(n).n^{1/2}\right]/A_{MCP}$$

Alternatively, 1/C can be calculated in a manner that uses different scales. Superimpose a grid on a map of a study area with cell size such that one cell contains all the locations of given animal. The area of the single can be considered as A and 1/C = n½/A. Now divide the single cell into four equal cells and calculate 1/C for each cell, letting A be the area of each new cell and n the number of locations in each new cell. The cells can be divided again each into four equal cells and the new 1/C calculated for each. In either of these approaches, a utility distribution can be calculated on different scales appropriate for different questions.

Gautestad and Mysterud also argued that the fractal approach to animal movements shows that "it is just as meaningless to calculate [home range] areas or perimeters as it is to calculate specific lengths of a rugged coastline." They concluded that home range areas cannot be measured because the number of data points needed for an accurate estimate exceeds the number that can be collected on most studies.

Unfortunately, Gautestad and Mysterud overstate their point. Clearly, home range boundaries and areas are simple and usually poor measures of animals' home ranges. The important aspects an animal's home range relate to the intensity of use and the importance of areas on the interior of the home range. So Gautestad and Mysterud are correct in playing down the importance of boundaries and areas.

Nonetheless, boundaries and areas can be estimated. Animals' home ranges have indistinct boundaries, just as the coastline of an island becomes indistinct when viewed using several different scales. But an island whose perimeter cannot be measured accurately nonetheless has a finite limit to its area, and that limit can be estimated.

Likewise, animals who confine their movements to local areas do have home ranges whose areas can be estimated, even if those areas must be estimated as a range between upper and lower limits, and even if the home range boundaries may never be known precisely. In addition, a useful estimate of the internal structure of a home range may be estimated with fewer data than needed to obtain reasonable estimates of the home range boundary or area.

In fact, during a finite period of time, an animal must confine its movements to a finite area and limits to that area can be estimated. Fixed kernel estimates of the areas of the annual home ranges of all bears located more than 300 times reached asymptotes after at most 300 chronological locations (131 ± 90, mean ±SD, n = 7; Powell, unpublished data; asymptote at 300 for a bear located more than 450 times, 95 percent home ranges).

However, equation 3.1 states that the estimated home range area must increase infinitely as the number of location data points used to estimate the home range increases. Clearly, this is a contradiction. The solution to the contradiction lies, with whether one includes unused areas within an animal's home range and whether one uses stable measures of the interiors of home ranges or uses unstable measures of the periphery.

Gautestad and Mysterud appear to have run their simulations using simulated utility distributions so large that their simulated animals could not use their whole "home ranges" within biological meaningful time periods. When this is the case, estimates of home ranges should increase in size as more and more simulated data points are used for the estimates. Indeed, after thousands of data points were used, the estimated home range areas do reach asymptotes at the areas of the utility distributions (Gautestad and Mysterud, personal communication), but note that this implies that equation 3.1 is not accurate for large n.

Some real animals may not use within a single year (or within some other biologically meaningful period) all the areas with which they are familiar. This raises the question of whether areas not used by an animal during a biologically meaningful period of time should be included in the estimate of its home range.

Perhaps Gautestad and Mysterud's simulated utility distributions actually represent animals' cognitive maps. Is an animal's cognitive map its home range? Or is its home range only the areas with which it is familiar and that it uses? No definitive answers exist for these questions. Equation 3.1 may be true for some animals. It is most likely to be true for animals that are familiar with areas far larger than they can use in a biologically meaningful period of time.

And if equation 3.1 is true, then the time periods over which we estimate home ranges may be as important as the numbers of locations. The time periods must be biological meaningful periods. To obtain accurate estimates of animals' home ranges, we may need to collect as many data as possible, organized into biologically meaningful time periods.

Another solution exists to the contradiction (not necessarily an independent solution). Gautestad and Mysterud estimated home range areas using 100 percent minimum convex polygons (but using the fudge factor Q(n)), which use only extreme, unstable data and must increase whenever an animal reaches a new extreme location.

They purposefully incorporated occasional sallies into their model but did not exclude them from their home range calculations. Small changes in sampling points at the extremes of animals' home ranges can lead to huge differences in calculated home range areas although the animals may not have changed use of the interiors of their home ranges. Home ranges areas for black bears using a kernel estimator, which emphasizes central tendencies, which are stable; home range estimates from kernel estimators do not change each time an animal explores a new extreme location.

Finally, Gautestad and Mysterud's model may be unrealistic. Any model of animal movement must be a simplification, so Gautestad and Mysterud's model does simplify animal movements. It does incorporate multiscale aspects of movement and appears to be a better model than, say, random walk models. Nonetheless, the multiscale random walk model still lacks important characteristics of true animal movements, and may thereby cause equation 3.1 to give a false prediction.

Even if equation 3.1 is false, the fractal utility distribution based on 1/C may still provide insight into use of space by animals. Unfortunately, by calculating C for each cell in a grid, one loses multiscale information that is available from an entire data set. In addition, 1/C provides no insight into estimated use of interstitial cells because it is only a transformation of the frequencies per cell ($n^{1/2}$ instead of n).

Finally, Vandermeer's cautions concerning grid dimensions must be addressed. One gains equal insight by calculating kernel home ranges and examining the probabilities for animals to be in cells of different sizes (scales), and kernel estimators are free of grid size constraints. Fractal approaches to animal movements may provide new insights into animals' home ranges, but their utility is still uncertain.

KERNEL ESTIMATORS

The best estimators available for estimating home ranges and home range utility distributions are kernel density estimators. Nonparametric statistical methods for estimating densities have been available since the early 1950s and one of the best known is the kernel density estimator. The kernel density estimator produces an unbiased density estimate directly from data and is not influenced by grid size or placement. Worton suggested that a kernel density estimator could be used to estimate home ranges of animals but little work had been published on the method as a home range estimator before Seaman's work, which is elaborated here.

Kernel estimators produce a utility distribution in a manner that can be visualized as follows. On an x—y plane representing a study area, cover each location estimate for an animal with a three-dimensional "hill", the kernel, whose volume is 1 and whose shape and width are chosen by the researcher.

The width of the kernel, called the band width (also called window width or h), and the kernel's shape might hypothetically be chosen using location error, the radius of an animal's perception, and other pertinent information. Luckily, kernel shape has little effect on the output of the kernel estimators, as long as the kernel is hill-shaped and rounded on top, not sharply peaked (deduced from criticisms by.

Gautestad and Mysterud, personal communication). Although no objective method exists at present to tie band width to biology or to location error, except that band width should be greater than location error, objective methods do exist for choosing a band width that is consistent with statistical properties of the data on animal locations.

Band width can be held constant for a data set (fixed kernel). Or band width can be varied (adaptive kernel) such that data points are covered with kernels of different widths ranging from low, broad kernels for widely spaced points to sharply peaked, narrow kernels for tightly packed points.

Although adaptive kernel density estimators have been expected, intuitively, to perform better than fixed kernel estimators, this has not been the case. The utility distribution is a surface resulting from the mean at each point of the values at that point for all kernels. In practice, a grid is superimposed on the data and the density is estimated at each grid intersection as the mean at that point of all kernels. The probability density function is calculated by multiplying the mean kernel value for each cell by the area of each cell.

Choosing band width is one of the most important and yet the most difficult aspects of developing a kernel estimator for animal home ranges. Narrow kernels reveal small-scale details in the data, and, consequently, tend also to highlight measurement error (telemetry error or trap placement, for example). Wide kernels smooth out sampling error but also hide local detail. The optimal band width is known for data that are approximately normal but, unfortunately, animal location data seldom approximate bivariate, normal distributions.

For distributions that are not normal, a band width more appropriate than that for a normal distribution can be chosen using least squares cross validation. This process chooses various band widths and selects the one that provides the minimum estimated error. Seaman found that cross-validation chooses band widths that estimate known utility distributions better than do band widths appropriate for bivariate normal distributions.

Using computer simulations and telemetry data for bears, Seaman explored the accuracy of both fixed and adaptive kernel home range estimators and compared their accuracies to the harmonic mean estimator.

He used simulated home ranges that looked much like real home ranges but he knew the utility distributions for the simulated home ranges. Seaman chose points randomly within the simulated home ranges, simulating the collection of telemetry or trapping or sighting location data, and then he estimated the simulated home ranges from the "location" data points.

He then compared the kernel estimators to the harmonic mean estimator because the harmonic mean estimator was widely used into the early 1990s, it appeared preferable to most well-known nonkernel estimators, and Seaman's comparisons can be extrapolated to other home range estimators through Boulanger and White's results. Seaman found that the different home range estimators varied greatly in accuracy of estimating both home range areas and utility distributions.

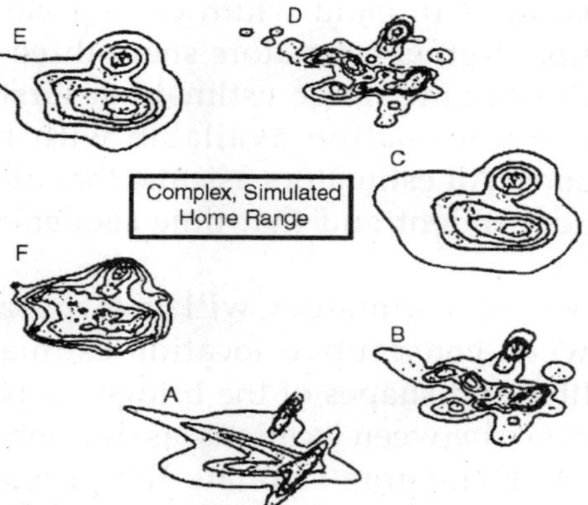

Fig. A complex, simulated home range. (A) True density contours. (B) Fixed kernel density estimate with cross-validated band width choice. (C) Adaptive kernel density estimate with crossvalidated band width choice. (D) Fixed kernel density estimate with ad hoc band width choice. (E) Adaptive kernel density estimate with ad hoc band width choice. (F) Harmonic mean estimate.

The fixed kernel estimator, using cross-validation to choose band width, yielded the most accurate estimates of home range areas and had the smallest variance. These estimates averaged 0.7 percent smaller than the true areas of the simulated home ranges, whereas the adaptive kernel estimates averaged about 25 percent larger than true. The harmonic mean estimator overestimated true home range area by about 20 percent. The cross-validated, fixed kernel estimator also estimated the shapes of the utility distributions the best.

The adaptive kernel estimators performed slightly worse than the fixed kernel estimators in all of the tests, apparently through overestimation of peripheral use. Adaptive kernel estimators also appear sensitive to autocorrelation within data sets. The amount of kernel variation can be adjusted for adaptive kernel estimators, but Seaman has found no consistent or predictable pattern of adjustment that minimizes error for these estimators.

Consequently, the best estimators at present are fixed

kernel estimators with band width chosen via least-squares cross-validation. Kernel estimators share three shortcomings with most other home range estimators. First, they ignore time sequence information available with most data on animal locations. All estimators assume that all location data points are independent and that time sequence information is irrelevant.

Future kernel estimators will incorporate brownian bridges between consecutive location estimates, with the heights, widths, and shapes of the bridges dependent on the time and distance between locations, as developed by Bullard. Second, kernel estimators estimate the probability that an animal will be in any part of its home range; therefore, they sometimes produce 95 percent home range outlines that have convoluted shapes or disjunct islands of use.

For example, the 95 percent fixed kernel home range for an adult female black bear, bear 61. Bear 61's home range in 1983 nearly surrounds a large area not designated as her home range. Surely, this bear was familiar with the surrounded area and included it on her cognitive map; however, she chose not to use that area regularly in 1983. In other years, she did use that area. The fixed kernel estimate of bear 61's home range accurately quantifies the area that she used in 1983 but may not accurately define the area that she actually considered to be her home range in 1983.

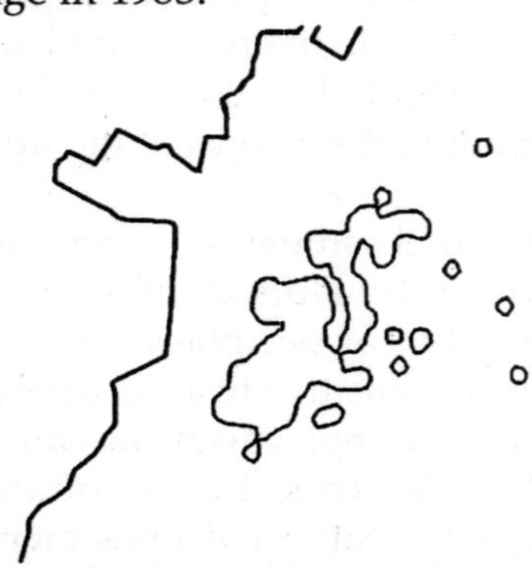

Fig. The 95% Fixed Kernel Home Range for Adult Female Blac K Bear 61 in 1983.

Bear 61's home range nearly surrounds a large, central area not designated as her home range. The thin dotted line marks the study area border.

Third, related to the second problem, kernel estimators estimate the probability that an animal will be in any part of its home range but do not estimate how important that part of the home range is to the animal. For researchers asking questions related to time or studying animals for whom time and importance coincide, no problem exists because home range estimators provide probabilities for use of space or, alternatively, probabilities for extent of time in given areas.

This aspect of home range estimators is a problem for researchers interested in the underlying importance of habitats or landscape characteristics when time and importance may not coincide. Some parts of a home range that are used little may be very important because they contain a limiting resource needed only at low levels of use. If time and importance do not coincide, kernel estimators (and all other estimators) do not estimate importance accurately.

HOME RANGE CORE

Particular parts of an animal's home range must be more important than other parts. In general, foods and other resources are patchily distributed, so the parts of a home range with greater density of critical resources ought, logically, to be more important than areas with few resources. For years, biologists have conceived the core as the part of an animal's home range that is most important to it. To understand home ranges well, identifying cores is important, if cores do indeed exist.

My understanding of a home range core has two major parts. First, a core must be used more heavily than the apparent clumps of heavy use that occur from uniform random use of space within a home range. Random use of space leads to some areas being used more than others, even though no place is more important to the animal than any other. Therefore, random use of space leads to apparent clumps of use in some places and little use of other places.

Consequently, the core of a home range must be used more than expected by random use, which means that for a home range to have a core, use of space within that home range must be statistically clumped and not random or even. Testing for clumped versus random (or for clumped versus even) use of space is usually straightforward.

In a uniform random distribution, the mean equals the variance. If an animal uses space at random, then the mean number of locations in each cell in its home range equals the variance. If the variance is significantly greater than the mean, then use is clumped; if the variance is less than the mean, then use is even. Note that many nonrandom distributions have means equal to their variances. Therefore, equal mean and variance does not prove uniform random use of space, but unequal mean and variance does disprove random use of space.

Second, a core must not be strictly determined by home range area. Animals with home ranges of equal size but with different patterns of home range use (e.g., central place foragers, strongly territorial animals, extensive wanderers) should have differently sized cores. Any technique developed to identify the core of a home range must reflect this biological understanding of what a core is.

Most definitions of cores have been ad hoc or subjective. Many define the core as the smallest area with an arbitrary probability of use (e.g., the smallest area enclosing 25 percent of total use). A crucial problem with such definitions, beyond their subjectivity, is that the cores so defined are not strictly tied to intensity of use of space.

In fact, animals that use their home ranges randomly or in an even fashion have cores by these definitions. Samuel et al. developed an objective method for identifying the maximum possible core: all parts of the home range used more heavily than would result from evenly distributed use. Although this definition is objective, it is still arbitrary, it allows cores to incorporate space used little more than adjacent space, and it defines cores for animals that use space randomly.

Seaman introduced a technique for identifying cores that is objective, not arbitrary, and that allows the animals themselves to define their cores. Our technique is based on the logic used to identify behaviour bouts. Bingham and Noon used the same method and Harris et al. may have, but their explanation is not clear.

If an animal's use of space is random within its home range, a plot of home range area at a certain percentage use versus probability of use should yield a straight line going from the 100 percent home range at probability of use equal to 0 (100 percent of the home range has probability of use 0 or above) to the 0 percent home range at probability of use greater than the maximum probability of use.

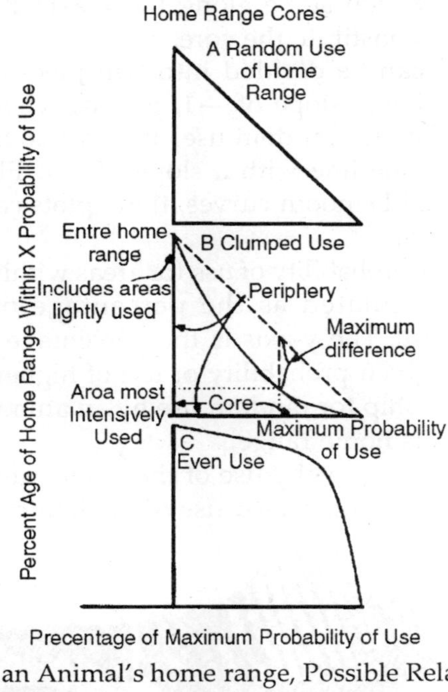

Fig. For an Animal's home range, Possible Relationships between Probability of use and Percentage of Home Range with the Probability of use or Greater.

If probability of use is transformed to percentage of largest probability of use, then both x and y axes on the graph range from 0 to 100 and the descending line representing random use has a slope of −1. If use of space by an animal is clumped, however, its curve will sag below the line of random use and if use of space is even (all areas used with equal intensity), the graph will remain as a high plateau from x equals 0 probability of use up to x equal some large probability of use and then plummet.

When use of space by animals is clumped, Seaman defined as an animal's core the areas in its home range that are used most intensively. The parts of the home range mapped onto the steeply descending slope of the area—probability curve along the y-axis are used least and constitute the periphery of the home range. The parts of the home range mapped onto the nearly horizontal slope along the x-axis are used most intensively and constitute the core.

The curve can be divided into two pieces at the point whose tangent has a slope of −1, that is, whose tangent is parallel to the line for random use. This is also the point that is furthest from the line with a slope of −1. Plots of actual data may not yield smooth curves; these plots can be fit with reasonable curves.

The x-axis is probability of use for areas within an animal's home range calculated as the percentage of maximum probability of use. The y-axis is the percentage of the home range with the given probability of use of higher use.

- Relationship for random use by an animal of the area within its home range,
- Clumped or patchy use of the home range,
- Even or overdispersed use of the home range.

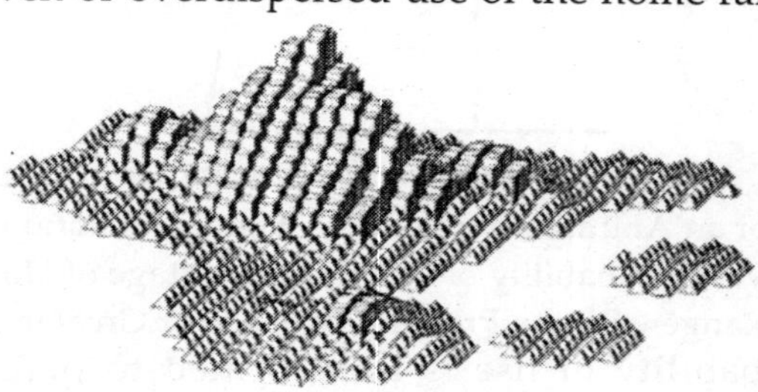

Fig. Core area and home range for an adult female bear. The core is shown with flat-topped symbols, the periphery with triangular symbols.

This criterion for a home range core clearly identifies the most intensively used areas within an animal's home range, and it allows the data to decide where the boundary between core and periphery should be located. The criterion is objective and, for me, intuitive. By this objective, each animal's core, if it has one, is at a different probability of use. In addition, some animals have large cores and some have small cores.

QUANTIFYING HOME RANGE OVERLAP AND TERRITORIALITY

Home ranges of conspecifics often overlap, sometimes extensively. For a population, hypotheses can be tested regarding simultaneous use of areas of home range overlap. Subsets of a population may exhibit different patterns of simultaneous use. Relatives, for example, may use their areas of overlap more than expected from random use, whereas nonrelatives may avoid each other and use areas of overlap less than expected. Intrasexually territorial animals may spend less time in their areas of overlap with members of the opposite sex than expected.

For some species, territorial behaviour has been documented objectively. Extensive experimentation with limiting resources and territorial displays and calls has defined territoriality clearly in many birds. For many species, however, apparent lack of overlap of home ranges is the only evidence of territorial behaviour.

For many carnivorous mammals, territory defence and responses to scent marks are difficult to document. For such species, home range overlap can be quantified in an objective manner that weighs probability of use of different parts of a home range or territory. Home range overlap can then be compared statistically among populations or species that appear to differ in territorial behaviour but for whom territorial behaviour has not been manipulated experimentally.

Doncaster defined two types of overlap, called static and dynamic interactions. Static interaction is the spatial overlap of two home ranges and dynamic interaction involves interdependent movements of the two animals whose home ranges overlap. These types of overlap can be quantified in several different ways.

STATIC INTERACTIONS

Area of overlap is a poor estimate of the effect or importance of the overlap on two individuals whose home ranges overlap. Areas of overlap vary in probability of use and the two individuals may have a large overlap of areas used little by each or a small overlap of areas used intensively by each. Although Genovesi and Boitani (1997) found that area overlap of minimum convex polygons correlated strongly with weighted overlap, this need not be the case for all populations or individuals.

Such probability density functions could be generated by a fixed kernel home range estimator or by any other estimator that produces accurate utility distributions. The first, I_p, is

$$I_P = \sum_{k \in 0} P_{ki} \cdot \sum_{k \in 0} P_{kj}$$

where p ki and p kj are the independent probabilities that at any arbitrary time animals i and j are in cell k of a study area that has a grid superimposed on it and that cell k is within the area of overlap, O, of the animals' home ranges. I_p ranges from 0 to 1. This index is the simple probability that the two animals i and j will be in their area of home range overlap at the same time were they to move independently of each other. Of course, most animals do not move without respect to the movements of other animals. Consequently, static interactions should be studied in conjunction with dynamic interactions.

A similar index, I L, is Lloyd's

$\overset{*}{m}$ index of mean crowding, which Hurlbert identified as probably the least biased overlap index:

$$\overset{*}{m} = N \cdot \sum_{k} P_{ki} \cdot P_{kj}$$

where N is the number of cells in which animal i or animal j (or both) has nonzero probability of use. The mean crowding index standardizes overlap by the probability of both animals being in the same cell at the same time were each to use all cells in the combined home ranges evenly and without respect to use by the other.

Seaman indexed the abilities of different home range estimators to reconstruct a known utility distribution by calculating the summed squared differences between predicted and known values for each cell. This index can be used to index overlap of two home ranges as Ó(p ki " p kj)2, where p ki and p kj are defined as above, and emphasizes differences where probability of use is high. The behaviour of this index has not been explored.

Doncaster indexed overlap using Spearman's coefficient of rank correlation. Spearman's r is calculated for the utility distributions of two animals with overlapping home ranges, or for the frequencies of use of cells in a grid. Doncaster showed that the index behaves well and that nonlinear responses of the index outside of the area of overlap (where one individual has probability of use equal to 0) do not affect the overall usefulness of the index. This appears to be a robust index that can be used broadly to index overlap of home ranges.

I index pairwise overlap only for animals whose home ranges overlap, or are adjacent, in the same year, that is, animals for which $I_p > 0$. This criterion sometimes excludes from analyses two home ranges that are adjacent but do not overlap, possibly because of sampling error. An objective method of choosing nearest neighbors is needed.

Because an animal's home range can overlap with the home ranges of several other animals, all pairwise index values within a study site are not strictly independent. Similarly, for studies that follow some individuals for more than 1 year, index values for different years may not be independent. Statistical tests must be controlled for both individuals in each pairwise overlap when testing for differences among sites, among years, or among populations of different species.

These indices of overlap and similar indices can be used to compare overlap among sites, to compare changes in overlap with changes in food or other limiting resources, or to deduce territorial behaviour when active defence, scent marks, or calls have not been documented.

DYNAMIC INTERACTIONS

Several approaches can be used to quantify and test whether two individuals affect the behaviour of each other. The indices test predominantly for attraction or avoidance.

Doncaster, Horner and Powell and Minta used chi-square and G tests to explore whether two animals located at the same time tended to be found together. In a 2 × 2 contingency table, paired and unpaired distances can be labeled as near (animals together or associated) or far (not together or not associated). Doncaster labeled the foxes he studied as near when they were close enough to detect each other; Horner and Powell labeled black bears as near when they were closer than the median distance of telemetry error.

Other objective criteria might be based on the animals being within an area of overlap at the same time. The N paired locations are a sample estimating how often the two animals are close together. By taking each location for each animal and calculating the distance to the N — 1 locations of the other animal not taken at the same time, one obtains a sample of N2 — N distances that can also be divided into near and far and estimate how often the animals would be found near each other if the movements of each were unaffected by the other. A significant chi-square or G value indicates that the animals attract or avoid each other. Minta noted that chi-square statistics behave better than G with small sample sizes.

Kernel estimates of utility distributions to estimate the probabilities that two animals would be in their area of overlap simultaneously. Because p ki is the probability that at any given time individual i will be found in cell k, L p is the probability that two animals are in their area of home range overlap at the same time if each uses the area without regard to use by the other. L p can be tested against the actual proportion of time (proportion of location estimates) that either animal spends in the area of overlap with the other.

Because the two animals are unlikely to be located the

same number of times, but must be in the overlap together the same number of times, the proportion of locations in the area of overlap will differ slightly between the two. This approach cannot be used to test whether two specific individuals attract or avoid each other but can be used to test whether classes of animals exhibit attraction or avoidance.

Minta developed further tests for attraction or avoidance to an area of overlap that allow researchers to test for more specific use of areas of overlap and that accommodate more diverse data than needed for the approaches used by Doncaster and my coworkers. Minta showed how to test whether one animal of a pair is attracted but the other not, how to test for attraction when animals are not always located simultaneously, and how to test for attraction when home ranges are not known but the area of overlap is.

TESTING FOR TERRITORIALITY

For many animals, territory defence is difficult or impossible to document but patterns of home range overlap can be documented. Such patterns of overlap can often be used to deduce territorial behaviour. Table gives overlap index results for territorial wolves in northeastern Minnesota (Mech, personal communication, unpublished data), for adult female black bears in North Carolina and in Minnesota that appear to differ in territorial behaviour, and for intrasexually territorial stone martens (Martes foina) in Italy.

Home ranges of wolves in different packs, and pack home ranges (from combined home ranges of pack members), overlapped little compared to home ranges of wolves in the same pack. Wolves in the same pack have home ranges that overlap extensively and Spearman's r is positive, showing significant correlation of the home ranges of wolves in the same pack and attraction among wolves of the same pack.

Spearman's r is negative for wolves of different packs and for pack home ranges, indicating avoidance by wolves of different packs. Such quantified patterns of home range overlap and lack of overlap agree with the extensive field observations that wolves defend territories.

Overlap of home ranges for bears in the North Carolina site was significantly greater than was overlap for bears at the Minnesota site. Productivity of food was significantly higher for the North Carolina population and anecdotal evidence indicates that the lack of home range overlap for the bears in Minnesota was caused by territory defence. No territorial behaviour was documented in North Carolina, but bears in both sites were seldom seen by the researchers.

Table provides solid evidence for differences in home range overlap and for differences in tolerance between individual adult females at sites differing in productivity of food. The difference in home range overlap between the two sites is consistent with observations of territorial behaviour and with territory theory. Nonetheless, black bears in North Carolina have negative values of Spearman's r, indicating that avoidance does occur among these bears at some scale.

Home range overlap differed significantly between stone martens of the same sex and martens of opposite sex using two of the three indices for overlap. Intrasexual territoriality has been deduced from observations of home range overlap, but objectively quantified overlap that incorporates probability of use has been lacking. Erlinge suggested that males of Mustela species avoided females' territories, indicating that large overlap in space may not indicate large overlap of important space.

The data for stone martens indicate that overlap between the sexes is significant. That Spearman's r is negative both for members of the same sex and for members of opposite sex indicates that members of opposite sex also exhibit some avoidance but not as much as do members of the same sex. Values of Spearman's r vary as the size of the cells in the grid varies and the mean difference in r between the sexes is maximal at an intermediate cell size.

When a cell size larger than the average home range area is used, home range overlap is similar within and between sexes. For cell size smaller than home range area, overlap among members of the same sex becomes smaller than overlap between the sexes. When cell size is much smaller than mean home range area, overlap within and between the sexes again becomes similar, indicating that members of the opposite sex with overlapping home ranges avoid each other on a fine scale. Taber et al. found similar avoidance by groups of Chacoan (Catagonus wagneri) and collared peccaries (Tayassu tajacu).

I_p is little affected by cell size as long as cell size is small compared to home range size because probabilities of being in cells are summed over the area of overlap. Lloyd's index $\overset{*}{m}$ and Spearman's r are both sensitive to cell size and therefore have the potential to give different results with different cell sizes. Cell size must match the grain or scale of the hypotheses being tested. Therefore, I_p shows that stone martens of the same and opposite sex show different patterns of home range overlap, whereas Spearman's r is able to show that spatial avoidance may occur on a scale much finer than that of individual home ranges.

LESSONS

Within Burt's definition of home range, tremendous latitude exists for different approaches to understanding what a home range is. No consensus exists for a single, precise definition of home range and this lack of consensus leads to some of the confusion regarding home ranges. Many researchers may be unaware of the confusion. The major confusion appears to be between those who use a conceptual definition and those who use an operational definition.

Burt's definition is conceptual and has two components: familiarity and use. Operational definitions and home range estimators quantify use and may add some estimate of familiarity. The minimum convex polygon estimator assumes that animals are familiar with all areas within the extremes of their movements but not familiar with any area outside those extremes, no matter how close.

Kernel estimators assume that animals are familiar with, and have interest in, areas surrounding their movements in a fashion that decreases as does the shape of the kernels. The only quantification of a home range that excludes some guess about familiarity would include only areas over which an animal actually traveled.

Should areas to which an animal seldom travels but with which the animal is familiar be included in its home range? This question addresses the difference between an animal's home range and its cognitive map. An animal's cognitive map includes all areas for which the animal has information, whereas its home range to include only the areas with which it maintains familiarity.

Familiarity may be maintained through the senses, without actual visits. In addition, familiarity is undoubtedly both graded and of a multiscale nature, making familiarity extremely difficult to quantify. How unused areas are quantified must be considered by researchers. To some extent, how unused areas are quantified depends on research questions. However, researchers must understand how home range estimators estimate familiarity and must use estimators that are consistent with their own concepts of home range and with their research questions. No agreement exists as to how best to quantify familiarity.

What currencies are best for quantifying home ranges and territories? Home ranges should not always be quantified with respect to the time that an animal spends, or is predicted to spend, in different places. Most researchers use time to index importance, but what really counts is fitness. For some questions, home ranges may need to be quantified by probability density functions of energy expenditure or energy acquisition. Ultimately, for much research, home ranges should be quantified as probability density functions for contributions to fitness. How does a person map a home range's contributions to fitness?

A strong correlation between home range size and the size of the periphery for black bears, suggesting that peripheral parts of home ranges should receive more research attention than they do. Peripheries expand and contract as home range cores contain or fail to contain necessary resources. How are peripheries expanded? How does an animal choose where and how to expand its peripheral home range? Does the expansion depend on past knowledge and, if so, how did the animal obtain that knowledge? How important are occasional sallies? And if occasional sallies are important, should they be included in home ranges? Conversely, what critical resources are contained in a home range core?

How can estimates of animals' home ranges be improved? No universally best method can be developed for estimating home ranges of all animals because the best method for one research project may be different than that for another. Optimal methods depend on the hypotheses being tested and different hypotheses demand different analyses. Researchers must understand the strengths and limitations of each home range estimator and choose the one that provides the most accurate information related to the hypotheses being tested or to research objectives. Nonetheless, some generalities exist for improvement.

Home ranges must be quantified over biologically meaningful periods of time. Data collected over too short a time period may not sample an animal within all areas that it considers to be its home range. Data collected over too long a time period may suffer from changes in an animal's home range during the time period.

Kernels and brownian bridges incorporating time sequence information (i.e., in what direction is an animal most likely to travel next?) promise to be useful for many studies. Future kernel estimators should also use information on telemetry error or area of trap attraction and information on an animal's radius of perception when calculating band width. Using probabilities generated by kernel estimators to explore use of space at different scales has much promise.

The fractal estimator, 1/C, may have promise. A future fractal estimator could become an estimator of choice for some research if development allows C for each cell to include time and multiscale information from the entire data set (not just from one cell), finds a way to determine optimal cell size for research questions of known scale, and solves the critical problem of C depending on an unstable estimate of home range area.

All methods of estimating and quantifying animals' home ranges have problems and no method is best for all research. Because of the myriad problems associated with quantifying home ranges, especially boundaries and areas of home ranges, White and Garrott suggested that the home range concept is obsolete and that our understanding of animal ecology and behaviour will advance best through research directed toward unambiguous analyses of data.

Gautestad and Mysterud implied the same when they argued that boundaries of home ranges cannot be quantified precisely. Clearly, as White and Garrott stated, "home range estimates are a poor substitute for good experimental protocol," but home range estimates and good research protocol are not mutually exclusive.

Animals do have home ranges. They do not move randomly through the world but stay in confined, local areas for days, seasons, or even years. To understand why animals live where they do, why they go certain places to do certain things, and how they share or divide the locale, researchers must grapple with the concept of home range.

For some questions, researchers should do as White and Garrott and as Gautestad and Mysterud suggested: Document exact animal locations, distances moved and rates of movement between locations, and so forth.

Answering other questions will require estimates of home ranges, especially estimates of how animals use space within their home ranges. Exploring the concept of home range will improve our understanding of how animals conceive and perceive where they live and will further our understanding of animals' cognitive maps of the land in which they live. We have a sense of place for where we live; other animals do, too. That is what we seek to understand.

Chapter 3

Habitat Evaluation in Delusions

Management of wildlife populations, whether to support a harvest, conserve threatened species, or promote biodiversity, generally entails habitat management. Habitat management presupposes some understanding of species' needs. To assess a species' needs, researchers commonly study habitat use and, based on the results, infer selection and preference. Presumably, species should reproduce or survive better (i.e., their fitness should be higher) in habitats that they tend to prefer.

Thus, once habitats can be ordered by their relative preference, they can be evaluated as to their relative importance in terms of fitness. Managers can then manipulate landscapes to contain more high-quality habitats and thus produce more of the targeted species. Habitat manipulations specifically intended to produce more animals have been conducted since at least the days of Kublai Khan.

However, the processes of habitat evaluation are fraught with problems. Some problems are specific to the methods used in the data collection or analyses. Many of these problems have already been recognized, and discussions about them in the literature have prompted a host of evolving techniques.

Other problems are inherent in the two most basic assumptions of this approach: that researchers can discern habitat selection or preference from observations of habitat use and that such selection, perceived or real, relates to fitness and hence to population growth rate.

My goal is to illuminate the scope of the problems involved in habitat evaluation. Assessments of habitat selection and presumed importance are done so often, and study methods have become so routine, that it is apparent that researchers and managers tend to believe that the major problems have, for the most part, been overcome.

TERMINOLOGY

The word habitat has two distinct usages. The true dictionary definition is the type of place where an animal normally lives or, more specifically, the collection of resources and conditions necessary for its occupancy. Following this definition, habitat is organism specific (e.g., deer habitat, grouse habitat). A second definition is a set of specific environmental features that, for terrestrial animals, is often equated to a plant community, vegetative association, or cover type (e.g., deer use different habitats or habitat types in summer and winter).

Nonhabitat could mean either the converse of habitat in the first sense (a setting that an animal does not normally occupy) or the second (a specific vegetative type that the animal views as unsuitable); here, the two meanings of habitat converge.

Hall et al. argue that only the first definition of habitat is correct and that the second represents a confusing misuse of the term. They reviewed 50 articles dealing with wildlife—habitat relationships and, based on their definition, found that 82% discussed habitat vaguely or incorrectly. The prevalent use of habitat to mean habitat type, this alternative definition is legitimate and well understood in the wildlife literature.

Moreover, this common usage of the term is consistent with the normally accepted meaning of habitat use: the extent to which different vegetative associations are used. Hall et al. define habitat use as "the way an animal uses ... a collection of physical and biological components (i.e., resources) in a habitat" (emphasis mine), which seems difficult to measure.

Habitat selection and preference are also more easily understood in terms of differential use of habitat types. Selection and preference are often used interchangeably in the wildlife literature; however, they have subtly different meanings. The distinction posed by Johnson, who defined selection as the process of choosing resources and preference as the likelihood of a resource being chosen if offered on an equal basis with others. Peek suggested that innate preferences exist even for resources not actually available.

Furthering this concept, Rosenzweig and Abramsky characterized preferred habitats as those that confer high fitness and would therefore support a high equilibrium density (in the absence of other confounding factors, such as competitors). Thus use results from selection, selection results from preference, and preference presumably results from resource-specific differential fitness.

In controlled experiments, preferences can be assessed directly by offering equal portions of different resources and observing choices that are made. In the wild, however, preferences must be inferred from patterns of observed use of environments with disparate, patchy, and often varying resources.

Generally, the purpose for determining preferences is to evaluate habitat quality or suitability, the ability of the habitat to sustain life and support population growth. Importance of a habitat is its quality relative to other habitats—its contribution to the sustenance of the population.

Assessments of habitat quality and importance (i.e., habitat evaluation) are thus based on the presumption that preference, and hence selection, are linked to fitness (reproduction and survival) and that preference can be gleaned from patterns of observed use.

Use of habitat is generally considered to be selective if the animal makes choices rather than wandering haphazardly through its environment. Typically, the disproportionate use of a habitat compared to its availability is taken as prima facie evidence of selection. Although technically resource availabilityu encompasses accessibility and procurability, these attributes are difficult to measure, so it seems reasonable to equate habitat availability with abundance (typically measured in terms of area), as is normally done in habitat selection studies.

A habitat that is used more than its availability is considered to be selected for. Conversely, a habitat that is used less than its availability is often referred to as being selected against, or even avoided. This is poor terminology, however, in that it suggests that the animal preferred not to be in that habitat at all, but occasionally just ended up there. Use that is proportional to availability is generally taken to be indicative of lack of selection, which is also unfortunate terminology, as illustrated by the following examples.

Consider an animal living in an area with only two habitats and using each in proportion to its availability; from this we might assume that the animal was not exhibiting habitat selection. However, unless the animal was a very low life form, it certainly made choices as to when it visited each habitat and what it did when it got there; anytime it made a choice, and either stayed or moved, it selected one habitat over the other.

Arguably, if one analyzed these movements on a short enough time scale, habitat use would be disproportionate to availability, enabling detection of habitat selection. As the time scale is shortened, though, the sheer physical constraint of moving between the two habitats (i.e., the distance between them) also affects their relative use.

On the flip side, imagine a dispersing animal attempting to traverse an area with no regard for habitat. If its route was frequently diverted by the presence of other, more dominant resident animals, living in their presumably preferred habitats, the disperser's movements would appear to reflect habitat selection (i.e., selection for habitats not preferred by the residents).

Indeed, one could reasonably assert that this represents true habitat selection as defined earlier, in that the disperser chose to avoid habitats with dominant conspecifics and thereby improved its chance of obtaining resources and not getting killed; however, one could also legitimately contend that the disperser was simply exhibiting avoidance of conspecifics, and used whatever cues, including markings, droppings, and possibly habitat characteristics, to do so.

These are not trivial complications, but rather examples of the intrinsic ambiguities associated with the application of these concepts. Terms such as selection and preference can be clearly defined, but not easily measured in the real world.

METHODS FOR EVALUATING HABITAT SELECTION[H]

Three general study designs have been used to infer habitat quality. The first, generally called the use—availability design, compares the proportion of time that an animal spends in each available habitat type (generally judged by the number of locations, or less commonly, by the distance traveled) to the relative area of each type. The second, the site attribute design, compares habitat characteristics of sites used by an animal to unused or random sites.

These two designs generate measures of selection for various habitats or habitat attributes, and habitat quality or importance is inferred from the magnitude of this apparent selection. The third method, which the demographic response design, uses a more direct approach for assessing habitat quality by comparing the demographics (density, reproduction, or survival) of animals living in different habitats. This design thus circumvents the need to interpret animal behaviour (habitat choices).

USE—AVAILABILITY DESIGN

Among studies of birds and mammals, the use availability design is the most popular. Habitat-related papers dealing with birds and mammals published in the Journal of Wildlife Management during 1985–1995 and found that most relied on a use—availability study design to assess habitat selection, preference, or quality.

Thomas and Taylor further categorized use—availability studies into three approaches: one in which habitat-use data are collected on animals that are not individually recognizable (e.g., visual sightings or sign), one in which data are collected on individuals (e.g., radiocollared animals) but habitat availability is considered the same for all individuals (so individuals are typically pooled for analysis), and one in which use and availability are measured and compared for each individual. They also reviewed papers published in the Journal of Wildlife Management and found that nearly twice as many studies collected data on individuals but pooled them for analysis than either of the other two approaches.

Studies that pooled animals for analysis have commonly compared frequencies of use and availability for an array of habitats using a chi-square test. Determination of which habitat types were used more or less than expected is generally made by comparing availability of each habitat type to Bonferroni confidence intervals around the percentage use of each type. This procedure was described initially by Neu et al. and clarified by Byers et al., although a more accurate method of constructing such confidence intervals was recently proposed by Cherry.

If the areas of available habitats are estimated (e.g., from sampling) rather than measured (e.g., from a map), use and availability should be compared with the chi-square test for homogeneity rather than goodness-of-fit. A chi-square goodness-of-fit test assumes that the availabilities are known constants against which use is compared, so if availabilities are actually estimated, with some sampling error, this test is more prone to indicate selection when there is none.

Various other methods of comparing use and availability have been advanced but less often used in wildlife habitat studies. Ivlev proposed an electivity index to measure relative selection of food items on a scale from "1 to 1; this has since been adopted for some habitat selection studies. However, Chesson noted that Ivlev's index may yield misleading results because it varies with availability even if preference is unchanged, and advocated use of a 0 to 1 index originally proposed by Manly et al., also for feeding preference studies.

This Manly—Chesson index is simply the proportional use divided by the proportional availability of each habitat, standardized so the values for all habitats sum to 1. As adapted to habitat studies, it is interpretable as the relative expected use of a habitat had all types been equally available (i.e., preference). Thus in an area with four habitats, an index of 0.25 for each habitat would indicate no preference, whereas deviations from this would indicate relative preference for or against certain habitat types.

Heisey and Manly et al. extended this method to test for differences in habitat preference among individuals or sex–age groups, and also showed how to test for statistically significant differences among preferences for different habitat types. Kincaid and Bryant and Kincaid et al. offered an alternative method that scores relative differences between use and availability for habitats defined as geometric vectors.

Most studies using these tests pooled data among individuals, so that animal captures, sightings, radiolocations, and so on represented the sample units. Aebischer et al. pointed out that this constitutes pseudoreplication and advised comparing use to availability for each animal individually (i.e., so individuals are the sample units).

Several methods have been developed specifically to do this. Of these, the most commonly used is Johnson's, which is based on the difference between the rankings of habitat use and the rankings of habitat availability. This method also provides a means of detecting statistically significant differences among habitats, not just a relative ordering of their selection. Moreover, because comparisons are made on an individual-animal basis, habitat availability can be considered either within each individual home range, or within the study area as a whole.

Johnson defined first-order selection as that which distinguishes the geographic distribution of a species, second-order selection as that which determines the composition of home ranges within a landscape, and thirdorder selection as the relative use of habitats within a home range. Thus, both second-order and third-order selection can be addressed with Johnson's technique; with chi-square tests it is possible but more difficult (because of sample size constraints) to consider both of these levels of selection.

Alldredge and Ratti compared four methods (including the chi-square, Johnson's, and two others based on individual-animal comparisons) in simulated conditions and found that none performed consistently better than the others. However, some methods are better suited for given situations. For example, because data for all animals are generally pooled for chi-square tests, unequal sampling among individuals could strongly affect the results if all individuals did not make similar selections.

Conversely, the methods that weight animals equally, regardless of the amount of data collected on each, may be subject to spurious results caused by small sample sizes and variability among individuals. McClean et al. used real data on young turkeys (Meleagris gallopavo), which have fairly narrow and well-known habitat requirements, to compare results of six analytical techniques for assessing habitat selection. In this case, the methods that treat individuals as sample units tended to be less apt to detect habitat selection.

Aebischer et al. offered what appears to be an improved procedure for comparing use with availability on an individual animal basis. This method (compositional analysis) has become increasingly popular because it enables assessment of both second-order and third-order selection and yields statistical comparisons (rankings) among habitats.

Additionally, because the data are arranged analogous to an ANOVA, in which between-group differences can be tested against within-group variation among individuals, it provides a means of testing for differences among study sites (e.g., with different habitats, different animal density, or different predators or competitors), seasons or years (e.g., with different food conditions), sex—age groups, or groups of animals with different reproductive outputs or different fates.

SITE ATTRIBUTE DESIGN

Site attribute studies differ from use—availability studies in that they measure a multitude of habitat-related variables at specific sites and attempt to identify the variables and the values of those variables that best characterize sites that are used (often for a specific activity). With this design, the dependent variable is not the amount of use (as with use—

availability studies) but simply whether each site was used or unused (or a random location with unknown use); the independent variables can be many and varied. Use—availability studies generally just deal with broad habitat types, or if more variables are considered, they are analyzed individually.

A site attribute design was used in 45 (29 percent) of the habitat selection studies. Of these, 28 were on birds and 17 on mammals. This design requires measurement of habitat variables at some defined site, usually one that serves some biological importance to the animal. Nest sites of birds are easily defined and biologically important, and hence are often the subject of studies of this nature.

Habitat characteristics of breeding territories, drumming sites, and roosting sites also have been investigated. Among mammals, studies have focused on characteristics of feeding sites, food storage sites, resting sites, shelters, wintering areas, or areas recolonized by an expanding population. Other studies have compared habitat characteristics of randomly located sites to sites where birds or mammals were observed, radiolocated, or known to have been from remaining sign.

The statistical procedures used in such studies vary. Most have used multivariate analyses to differentiate combinations of variables that tend to be associated with the used sites. Discriminant function analysis (DFA) is the most popular of these. Logistic regression is an alternative, and is especially useful when the data consist of both discrete and continuous variables or are related to site occupancy in a nonlinear fashion.

DEMOGRAPHIC RESPONSE DESIGN

Ideally, studies should identify relationships between habitat characteristics and the animal's fitness. Studies employing use—availability and site attribute designs assume that certain habitat features are selected because they improve fitness. Demographic response designs attempt to test this more directly. However, refer to the measured demographic parameters in these studies as response variables, they really only represent correlates with given habitats.

Most of these investigated differences in animal density among habitats. Fourteen studies, all on birds, related reproduction (i.e., nesting success) to habitat of nest sites. Three studies, two on birds and one on mammals, attempted to find an association between habitat and survival, but only one detected such a relationship.

DEFINING HABITATS

The first prerequisite for assessing habitat selection is that habitats be defined as discrete entities. For use—availability studies in particular, the defined number of habitats can directly affect the results. Yet habitat distinctions often are not clear-cut. A researcher might distinguish two general forest types, uplands and lowlands, or might classify habitats by dominant overstory, or might divide these further by stand age or understory, and so on.

As more types are defined, sample sizes are reduced for observed use of each type, thereby diminishing the power of the statistical tests to distinguish differences between use and availability. Also, because the proportional use and availability of all habitats each sum to 1, the number of habitats distinguished affects all of these proportions. Aebischer et al. observed that this unit—sum constraint renders invalid many of the statistical tests often employed to compare use and availability because the proportions are not independent. That is, if one habitat type.

Has a low proportional use, others will have a correspondingly high use, and if there are only a few types, then the infrequent use of one type will lead to the apparent selection for another. Aebischer et al.'s method of compositional analysis was developed specifically to circumvent this problem.

Not just the number of types, but the criteria used to partition types may greatly affect results. Knight and Morris were able to visually differentiate 13 habitat types on landscape photographs of their study area, but postulated that only two broad classifications were distinguished by red-backed voles (Clethrionomys gapperi), the subject of their study. After analysis of their data, however, it became clear that from the voles' perspective, at least three functional habitats existed.

Another problem is the scale at which habitats are viewed. For example, an animal might appear to select for a certain habitat type, defined by a dominant cover type, whereas in reality it selected for certain specific kinds of sites that just happened to occur more commonly in that cover type than in others. An animal's choice of habitat type is often called macrohabitat selection and the choice of specific sites or patches within habitats is called microhabitat selection.

These may be perfectly hierarchical in that the most preferred microhabitats always occur within the same macrohabitat, in which case an animal may really select initially at the scale of macrohabitat, and then focus on specific sites within it. Schaefer and Messier observed this sort of nested hierarchy across a range of scales for foraging muskoxen (Ovibos moschatus) in the Canadian High Arctic.

Alternatively, the distribution of preferred microhabitats could be largely unrelated to the broader habitats defined by the biologist; in this case, a site attribute study might identify characteristics related to preferred microhabitats, whereas a use—availability study would detect no selection at the level of habitat type. This situation was apparently the case for wood mice (Apodemus sylvaticus) inhabiting arable lands in Great Britain: The mice seemed not to select (based on a use—availability study) from among three types of croplands (macrohabitats), but within each of these croplands they chose microhabitats with a high abundance of certain plants.

In sum, significant challenges in defining habitats include: partitioning them in terms of the features that the animals are selecting for, which are not necessarily the ones we most easily discern; delineating sufficient habitat categories to ensure that the truly important types are not lumped with and thus diluted by less important types; and not diminishing the power to discern selection by parceling out too many types.

MEASURING HABITAT USE

Sample bias is an obvious potential problem in measuring

habitat use. Interpretations of habitat use from visual observations of animals or their sign can vary among observers and sightability can vary among types of habitats, both of which can introduce biases in the data.

For example, Powell noted that fisher (Martes pennanti) tracks in snow were difficult to follow in habitats with dense vegetation, especially where fishers followed trails of snowshoe hares (Lepus americanus); in this case the bias against observing tracks in dense vegetation merely detracted from the overall conclusion that densely vegetated habitats were frequently used.

Counts of pellet groups (e.g., from ungulates or lagomorphs) may poorly reflect habitat use because defecation rates often vary with the food source, and hence the habitat type. Capture locations may be a poor indicator of habitat use because baits and other trap odors (e.g., from captures of other animals) may affect behaviors in an unpredictable way.

Telemetry also may yield biased data on habitat use because the detection of an animal's radio signal may depend on the habitat it is in, and location data obtained by triangulation have inherent associated errors. Intuitively, and as shown in computer simulations by White and Garrott errors in determining habitat use increase with increased habitat complexity and decreased precision in the telemetry system.

Errors do not necessarily introduce bias, but can if patch size differs among habitats (detected use would be underrepresented in habitat types that tend to occur as small patches) or if the animal preferentially used the edge of some habitat types but not others. Powell reported different perceptions of habitat use of fishers between his study, where he followed tracks in the snow, and another nearby radiotelemetry study; he attributed the difference to error in the telemetry system and consequent incorrect habitat categorization for animals near edges.

Nams showed that simply discarding locations because of large telemetry error, as is common practice, exacerbates bias; he offered a procedure for circumventing it, but few studies have used it. Kufeld et al. suggested using the habitat composition of error polygons formed by the triangulation of radio bearings, but this would not alleviate bias. Chapin et al. solved the problem a different way. In a study of habitat use of American martens (Martes americanus), which have a documented affinity for mid- to late-successional forests, they classified telemetry locations that were outside small patches of forest but within telemetry error of the edge as representing use of those forest patches.

Even if habitat use can be measured accurately, biases may result from sampling or analytical procedures. Habitat use may vary by individual, sex—age group, social status, time of day, season, and year, yet many (most) studies pool individuals and do not sample adequately. Schooley reviewed habitat studies published in the Journal of Wildlife Management and found that most lasted only 2 years, and most pooled results across years without testing for annual variation.

He used results of a black bear (Ursus americanus) study to show that habitat use can vary annually, and that the data pooled across years can yield misleading results. Beyer and Haufler found that most published studies that they reviewed collected data only during daylight hours; in their study of elk (Cervus elaphus), habitat use differed between day and night. Similarly.

Arthur and Schwartz reported diurnal and nocturnal differences in habitat use for brown bears (Ursus arctos) that fed at a salmon stream that was used by people during the day; this difference was detected with data from GPS collars, but was not apparent from conventional diurnal telemetry data. Ostfeld et al. and Belk et al. observed sex-related differences in habitats used by ground-dwelling rodents; Belk et al. remarked that combining the two sexes would produce a false perception of habitat use.

Paragi et al. observed differences in habitat use of resident and transient martens. Boitani et al. and Macdonald and Courtenay observed individual differences in habitat use, apparently related to social status. Bowers found that habitat use of eastern chipmunks (Tamias striatus) varied significantly with distance from their burrows, a finding noticeable only by considering the data on an individual basis. "It is time," Bowers commented, "that ecologists recognize that microhabitat selection and usage is a process involving individuals, not species."

Pooling individuals is common because sample sizes are typically too small to test for selection by individual. However, the statistical tests usually used assume independence among sample units, which is often not the case in studies that consider each location a sample. Some techniques consider animals as sample units, so lack of independence among locations within individuals is not problematic.

However, these methods are still subject to difficulties with lack of independence if animals are gregarious or territorial (social exclusion precludes use of certain habitats), or if the study subjects are related (habitat preferences possibly affected by a common learning experience) or are from the same social group.

In an effort to alleviate the problem of a lack of independence among individuals, Neu et al. used groups of moose (Alces alces) and Schaefer and Messier used herds of muskoxen as their sample units, rather than individual animals. Similarly, although Gionfriddo and Krausman monitored habitat use of individual radiocollared mountain sheep (Ovis canadensis), they considered groups of sheep their sample unit.

However, Millspaugh et al. contend that animals in a herd should be considered independent individuals if they congregate because of a resource rather than because of a biological dependence on each other. They provide a hypothetical example with elk, where 99 of 100 radiotagged animals congregated at a winter feeding area in one habitat and the remaining individual used a second habitat; at other times of the year the elk did not associate with each other. In this case, they argue that each radiotagged individual should be considered an independent sample.

In contrast, predators that hunt together in a pack and are thus dependent on one another cannot be considered to use habitats independently. Millspaugh et al. recommend tests to evaluate independence of habitat use by seemingly associated individuals.

MEASURING HABITAT AVAILABILITY

Measuring habitat availability is often more problematic than measuring use. Use—availability studies inherently assume that study animals have free and equal access to all habitats considered to be available. That is, at any given moment each study animal should be able to use any available habitat.

This assumption may hold if use and availability are measured for each animal individually. However, the assumption may be violated when animals are pooled for analysis and the available habitat is considered to be the same for all, yet some individuals may not even have all habitat types within their home range.

Johnson suggested that the habitat composition of home ranges compared with the habitat composition of some broader area should indicate the level of selection animals exercise when establishing their home range. Often the broader available area is considered to be that encompassed by the composite of the home ranges of all study animals. However, there are several problems with this.

First, animals cannot really select their ideal mix of habitats to compose their home range. Animals can only choose home range borders that encompass the best mix of habitats from what exists on the landscape; they cannot alter the mix to suit their needs. By analogy, a person may pick a town to live in, among several available, based on the resources available. One may also choose where to live within the town, but one cannot alter the layout of the town or the array of features available.

Second, animals may not have free and equal access to all areas when establishing their home range. Home ranges may be established near the natal area just because of familiarity with resources or neighboring animals, not any choice related to habitat composition. Analogously, people might remain in their home state or country not because they consciously chose it among all others, but because they never had the opportunity to visit other places, or because moving elsewhere, even if it seemed desirable in some respects, had too many costs.

Social constraints also may dictate choice of a home range by precluding access to certain areas. Extending the analogy with people, consider a house to be like a home range and a neighborhood a composite home range. The first few residents of a neighborhood might have selected where to live among houses that differed in various ways; however, as more people moved in, the choices narrowed, until no choice remained for the last resident.

If all houses were used, regardless of their quality, one could not discern after the fact which houses were preferred unless the "colonization" process was observed. Fretwell and Lucas proposed a corresponding model for animal populations. In an expanding population, preferred habitats are settled first, but as these are taken, animals are forced to settle in poorer and poorer areas.

However, unless they are strictly territorial, their ranges can overlap, so unlike the human example, they can choose to live in a preferred area even though another animal is already there. As animal density increases in the most preferred habitat, however, resources become less available to each individual, so the quality of the habitat from each resident's perspective diminishes.

Thus unless individuals benefit from the presence of others, their home range selection is negatively influenced by conspecific density. Other competing species have a further interacting effect on habitat availability and hence selection. Because competition changes each animal's perception of habitat availability, human measurement of availability, based on the assumption of free and equal access, is inevitably inaccurate. Animals tend to be more uniformly distributed across patchy landscapes than predicted from studies of habitat selection.

Another major problem in measuring habitat availability is the recognition and treatment of areas of nonhabitat that may exist within home ranges. Part of the difficulty arises simply because our concept of home range is too nebulous. Home range is typically defined as the area used by an animal for its normal activities, but home range area is a human perception, not a biological entity.

Humans may perceive the landscape as a mosaic of habitats that fit together like a jigsaw puzzle, on which are superimposed home ranges of animals. In contrast, animals may perceive the landscape as series of corridors or islands sprinkled in an ocean of nonhabitat.

If we unwittingly define available habitat from our human perspective, and include large patches of nonhabitat that the animal does not really perceive as among its choices of places to live, a comparison of use to availability might demonstrate nothing more than avoidance of the nonhabitat.

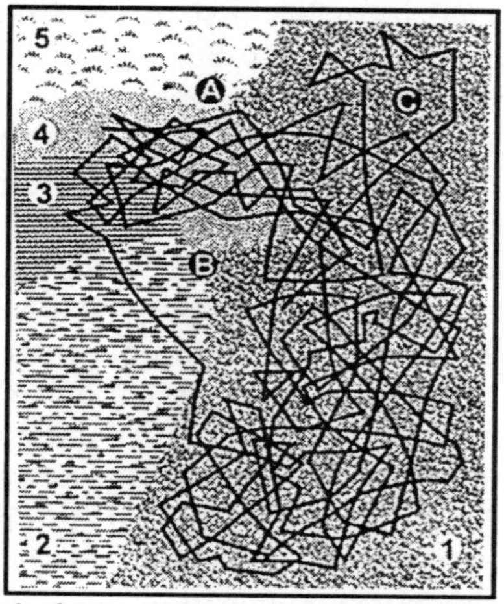

Fig. Hypothetical Movements of an Animal Overlaid on five (numbered) Habitat Types. Habitat Selection is often Assessed in Terms of Relative use Compared to Availability.

This would be grossly accurate, but not particularly insightful. An example was presented by Johnson, where mallards (Anas platyrhynchos) rarely used open water areas far from shore, but the area of open water was large. Standard means of comparing use to availability, such as the chi-square test, might show open water to be avoided and all other habitats selected; however, a knowledgeable duck biologist would recognize this as a trivial result, and might elect to exclude this obvious nonhabitat from the analysis.

Other cases may not be so clear-cut. Manly et al. presented an example with California quail (Callipepla californica), taken from a study by Stinnett and Klebenow. Bonferroni confidence limits, and hence perceptions of selection, depended on whether habitats that were not used as escape cover when the birds paired for mating were included or excluded from the analysis. In this case the habitats that were not used as escape cover during mating were not obvious nonhabitats because the birds used them in other circumstances and for other activities.

In this example, habitats 1, 3, and 4 were used and thus also available. Habitat 2 (depicted as a swamp) appears to have been traversed, possibly just to get from habitat 1 to habitat 3; if it was used simply because of its location, not because of its habitat-related attributes, a question arises as to whether it should be considered in the analysis. Conversely, although habitat 5 was not used, it may or may not be considered available.

Judged within the context of the home range boundaries, point A in habitat 5 appears to be unavailable, yet this point is closer to known locations of the animal than points B or C, which are both within the apparent home range. Habitat availability is a nebulous concept, and thus may be difficult to measure. Similarly, although the figure depicts a travel route, from which relative use of habitats might be deduced, most analyses deal with relative time, not distance, in each habitat (partly because telemetry data are generally comprised of point locations); it is unclear which is really a better measure of use.

An advantage of Johnson's technique is that the results are rather robust to inclusion or exclusion of habitat types that are rarely used. A problem with Johnson's technique is that because it is based on rankings of use and availability, habitats will not appear to be selected if their proportionate use is ranked the same as their availability.

Thus even if the animal spends an inordinate amount of time in the habitat that is most available, selection for this habitat will not be detected using this technique because both use and availability are ranked the same.

The Manly—Chesson index of habitat selection also does not fluctuate with inclusion or exclusion of seldom-used habitats, and Manly et al. showed that this index is much more versatile than Johnson's in many other respects. Recently, it was adapted by Arthur et al. to handle situations in which habitat availability changes. These authors recognized that habitats available to polar bears (Ursus maritimus) varied with changes in ice conditions and with movements of bears across their enormous home ranges.

Thus they defined availability separately for each radiolocation, using the habitat composition of a circle with a radius (from the radiolocation) equal to the expected distance a bear would travel during the time between radiolocations; habitat availability within these circles was then compared with the type of habitat the bear actually used the next time it was located.

Another attribute of Manly et al.'s procedure is that it can be used to analyse data from site attribute studies as well as use—availability studies, although site attribute studies also face problems in assessing availability. If used sites are compared to random sites, the universe from which the random sites are drawn must be defined.

As discussed earlier, that universe can be some arbitrarily defined study area, a composite home range of study animals, or each individual home range. Additional difficulties may arise if the comparison is between used and unused sites because errors may arise in distinguishing unused sites (i.e., nonobservation of use may not mean nonuse).

Furthermore, unused sites may be vacant for a variety of reasons, some of which are unrelated to the physical habitat (e.g., human disturbance, exploitation, predation, parasites, interspecific competition). Some predictive models have fared poorly when they did not consider such variables. Geffen et al. found, unexpectedly, that Blanford's foxes (Vulpes cana) in desert environments were rarely observed near springs, where water and food were most abundant, probably because this habitat was favored by and provided cover for potential predators.

In order to assess the criteria used by a species in selecting sites, investigators ideally should choose for comparison sites with both available resources and predators (or other confounding agents) present, as well as sites with only one or the other; however, such comparisons are unavailable in most field studies.

If a species is very selective in its choice of sites, differences between used and unused sites may be quite subtle; these subtleties would not be discernible in site attribute studies if the investigator chose unused or random sites that were very different from the used sites. The scale of comparison in this case would be too coarse. In an attempt to circumvent this difficulty, Capen et al. eliminated available sites in habitat types that were "radically different" from those that were used (analogous to eliminating nonhabitats in use—availability studies).

Conversely, if in attempting to use a finer scale of comparison one picked random sites from too narrow a universe, such that they were all very similar to the used sites, habitat differences might not be detected if a large portion of the random sites were used. This points out the advantage of distinguishing unused sites instead of just random sites and of selecting unused sites that are similar in many respects to the used sites.

Use—availability studies do not distinguish unused areas and so may be especially prone to problems of too fine or too coarse a scale of comparison. The coarse-scale problem (used and available areas are too dissimilar to detect the true basis for selection) may occur when composition of home ranges or habitat use within home ranges is compared to some broader study area. The fine-scale problem may occur when habitat use is compared to availability within home ranges.

Thus these two scales of comparison may yield different results. McClean et al. examined the effects of varying the definition of available habitat, from the entire study area to progressively smaller-sized circles around individual radiolocations. They found that selection became increasingly difficult to detect as availability was defined by a smaller and smaller area. This result is not surprising because the radiolocation represents use, so habitat composition within smaller areas around each location more closely matches areas of actual use.

ASSESSING HABITAT SELECTION[H]

Perceived habitat selection may vary with the technique chosen to compare use and availability or to compare attributes of used and unused (or available) sites. Some of this variation in perceived selection stems from the fact that different methods actually test different biological hypotheses and some is from the different assumptions inherent in these techniques and their sensitivity to violation of these assumptions.

Manly et al.'s technique can handle both use—availability and site attribute study designs. Moreover, it can be performed on an individual animal basis or with pooled data, it can be used to compare habitat selection among groups (e.g., species, sex—age classes, seasons, times of day, times within seasons), and it can incorporate both discrete and continuous variables. For these reasons, it has been heralded as a unified approach.

Manly et al.'s approach generates a resource selection probability function, giving the probability of a site being used as a function of various habitat variables. Each habitat variable can be tested to determine whether it contributes significantly to the probability of use. In the special case of only a single categorical habitat variable (i.e., habitat type), the function reduces to the Manly—Chesson selection index.

An advantage of this index, as discussed earlier, is that it is rather unaffected by the inclusion or exclusion of seldom-used habitats. In this sense, Chesson suggested that the index is a measure of preference that "does not change with density unless behaviour changes" and that it represents the expected use of the various resources if all were equally abundant.

Consider first the simple example presented by Chesson

to demonstrate the intuitiveness of the Manly—Chesson technique. The example deals with choice of foods. Suppose habitats A and B are equally available, and an animal spends 25 percent of its time in habitat A and 75 percent in habitat B.

Because the Manly—Chesson selection index represents the expected use when resources are equally available, the index for each habitat in this case simply equals their proportional use (0.25 and 0.75 for A and B, respectively). Now suppose that the same animal is placed in an area composed of 80 percent habitat C and 20 percent habitat B, and it uses C 40 percent of the time and B 60 percent.

The Manly—Chesson index would be 0.14 for habitat C and 0.86 for habitat B, suggesting that if habitats C and B had been equally available, they would have been used in these proportions. Because both A and C were compared against the same standard (habitat B), the results indicate that A would be preferred to C if those two types were offered together. However, given that the animal used A only 25 percent of the time but C 40 percent of the time, when in both cases the other choice was habitat B, the higher standardized selection index for A is not intuitive; these results are clearly a function of the higher availability of habitat C.

A fatal flaw of habitat selection studies in general, especially use—availability studies, is that they are based on the assumption that the more available a resource is, the more likely an animal should be to use it. This may not be true at all, may be true for only some resources, or may hold only within a narrow range of availabilities.

Manly et al. made the explicit assumption, applicable for all models that availability remains constant for the period of study (if availability changes seasonally, data can be analyzed by season). This may seem like a benign assumption, but in reality it masks a fundamental weakness of the process. Of what value are measures of selection if they are specific to a single array of habitats?

Measures of selection are supposed to be reflections of inherent preference—expected choices when availabilities of all habitat types are equal—so if selection appears to change as availability changes, then preference cannot be inferred from perceived selection when availabilities of habitats are unequal.

In other words, if the goal is to assess habitat preferences for a population of animals based on habitat selection observed among a collection of individuals in that population, then something is amiss if selectivity appears to differ among these individuals simply because they have different habitat compositions available to them.

Consider a human analogy that demonstrates the effects of changes in availability on perceived selection. While at home a person spends 50 percent of the time sleeping and 20 percent preparing food and eating meals in the kitchen; the bedroom occupies 20 percent of the area of the house, and the kitchen 10 percent.

Manly—Chesson selection indices for these rooms would be 0.51 and 0.40, respectively. Now suppose the person feels cramped in the kitchen and moves a wall, making it twice as big, at the expense of a room other than the bedroom. Afterwards the kitchen makes up 20 percent of the area of the house, the same as the bedroom, but use of the kitchen does not increase because it still takes the same amount of time to prepare and consume meals there.

The selection index for the kitchen thus drops to 0.25, despite the fact that it is now more comfortable and better serves its purpose. Moreover, although no changes were made to the bedroom, its selection index improved to 0.63 as a result of the renovations to the kitchen. Superficially, it would appear that the expense for remodeling was not worth it.

Analogously, one might imagine a situation in which an animal used a habitat substantially more than its availability, but used it only for sleeping. If that habitat became more available, the animal would not be expected to sleep more, so its selection for it would appear to decline. A management agency that produced more of this habitat because results of a habitat selection study showed it to be used disproportionate to its availability would be disappointed to find that these efforts made the animal's selection for it drop.

These examples demonstrate cases in which the activity

requires a fixed amount of time, so increasing availability of the preferred setting for that activity has no effect on how much time is spent there.

This situation is just a special case demonstrating the point that use and availability are not inexorably linked. In the example of the house, the renovated kitchen might entice the person to spend more time there, but only up to a point (one certainly would not sleep there).

Conversely, if the dining room had been remodeled at the expense of room in the kitchen, the person might not eat in the kitchen anymore, but, no matter how small it was, still prepare food there. Each room might thus have its own functional relationship between area and use. Similarly, if an animal prefers a certain habitat for resting because it offers protection from predators, it might spend more time resting in a larger patch of that habitat because it offers greater safety than a small patch.

Enlarging a patch that offers virtually no predator protection to a size yielding some predator protection might thus cause significantly increased use of the patch; however, additional enlargements might have progressively lesser effects on use because they do not add much predator protection, and eventually further enlargements do nothing, or might even attract a different predator, thus deterring use.

Various scenarios and corresponding relationships between patch size and use are plausible. Considering that the relationship between patch size and use probably varies among habitat types and the mathematical relationship between use and availability also differs among the various selection indices, it seems doubtful that one could assess selection just by comparing relative use to the relative area of different habitats.

Mysterud and Ims proposed a logistic regression model to compare use:availability ratios among study subjects that had differing habitat compositions available to them. This model thus provides a test of the assumption that use increases with increased habitat availability. Their method is applicable to cases in which habitats can be categorized into two discrete types (e.g., forested vs. nonforested, oak vs. nonoak). They reexamined two data sets that Aebischer et al. had analyzed using compositional analysis.

In one, use increased with increased availability of a habitat for 9 of 12 ring-necked pheasants (Phasianus colchicus); however, three individuals did not fit this trend. In the second example, gray squirrels (Sciurus carolinensis) showed an inverse relationship between use and availability of open habitats within their home ranges (the same unexpected relationship therefore existed for the alternate, forested habitat). It was surmised that size and interspersion of habitat patches greatly affected the choices that these animals made, more so than just total habitat area.

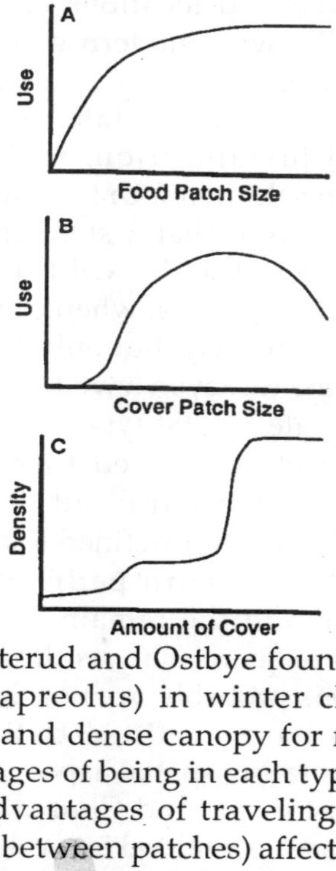

Similarly, Mysterud and Ostbye found that although roe deer (Capreolus capreolus) in winter chose open canopy habitat for feeding and dense canopy for resting, they had to balance the advantages of being in each type of habitat against the energetic disadvantages of traveling between them, so patch size (distance between patches) affected habitat selection.

Mysterud et al. suggested that for animals such as roe deer, which face tradeoffs in using different habitats, selection is not directly related to resource availability, so habitat rankings based simply on ratios of use to availability often are misleading. Bowyer et al. used a site attribute analysis to examine habitat selection related to various tradeoffs faced by black-tailed deer (Odocoileus hemionus).

Assessing selection can be extraordinarily complex because each habitat is not a single patch, but a series of patches of different sizes and shapes, each bordering other patches of different sizes, shapes, and habitat types.

Otis offered a model that tests for the disproportionate use of habitat types as well as habitat patches, thereby providing a means of assessing things such as minimum patch size requirements. Data for this model (patch size distributions for each habitat type and locations of animals in specific patches) are available with modern geographic information system (GIS) coverages.

This model still does not take into account habitat interspersion and juxtaposition, which probably have significant effects on selection for many species. For example, Porter and Church found that a standard use—availability analysis of habitat selection by wild turkeys indicated an avoidance of agricultural lands, when in reality turkeys used agricultural lands extensively, but only those near hardwood forests. In essence, the turkeys viewed the edge between field and forest as a separate habitat type.

Similarly, Neu et al. posited that moose might feed preferentially in a recent burn, but not too far from the surrounding forest. Thus they defined four habitat types—the interior of the burn, the burn periphery, the forest edge adjoining the burn, and the remainder of the forest—and through a simple chi-square analysis showed selection for the edge. Most situations probably are not this simple.

Many authors have admitted to the importance, but difficulty, of incorporating spatial aspects of habitats in use—availability analyses. Porter and Church proposed a method whereby the study area is gridded into cells and an assortment of habitat variables within those cells are examined through multivariate analyses to find those that best explain differential use of cells. Litvaitis et al. did just that in a study of bobcats (Felis rufus), which predated the paper by Porter and Church. Litvaitis et al. looked for associations (using regression and DFA) between the number of radiolocations within 25-ha cells inside home ranges and measurements of several habitat variables sampled there; however,

They found that these habitat variables poorly explained variation in frequency of use. Servheen and Lyon used a similar approach in assessing habitat selection by caribou (Rangifer tarandus). They measured habitat variables in 40-ha circles around telemetry locations and sought to find those that best differentiated the areas that the animals used seasonally.

Although they had no real measure of juxtaposition or interspersion of habitats, their 40-ha circles contained habitats neighboring the one actually occupied, so the composition of these circles gave an indication of habitat combinations that corresponded with seasonal use. In another similar approach, Clark et al. used grid cells that could encompass several habitat types near the locations of radiocollared black bears.

A suite of habitat characteristics (including the number of different habitat types) within each cell used by bears were combined to form what they called an ideal habitat profile. The habitat quality of each cell in the study area was then assessed by comparing it to this hypothetical ideal cell. Each of these studies looked at differential use, rather than use in terms of availability, and thus avoided the fatal flaw of habitat selection studies.

Site attribute studies are like the habitat use studies just discussed, except that instead of comparing cells with varying degrees of use, they categorize cells (sites) simply as used or unused; based on this, important habitat variables are identified. Interspersion and juxtaposition of habitats can thus be investigated. For example, Coker and Capen examined cowbird (Molothrus ater) selection for habitat patches of various size, shape, and location relative to other habitats by entering these variables in a logistic regression with use (used or not used) as the dependent variable.

Similarly, Chapin et al. compared habitat variables (including an index of the extent of habitat edge) in grid cells of different sizes that were used (i.e., had at least one telemetry location) by American martens with those in cells not used by martens, and also compared characteristics of forest patches that were used and not used.

McLellan argued that observed use is a better indicator of habitat selection than use relative to availability. He reasoned that an animal familiar with its home range knows the availability and location of resources, so an animal's location at any given moment represents selection.

He gave an example of a person at a buffet selecting a slice of beef from a 500-kg steer and an equalsized slice of pork from a 100-kg pig; based on use alone, pork and beef were selected equally, but compared to availability, pork appears to be selected over beef, which is obviously absurd.

However, had the steer and pig been cut up in equal-sized chunks and distributed over a large area, and after considerable searching the person still returned with an equal quantity of the two foods, active selection for pork would indeed seem apparent.

The key difference is that in the latter case the person had to search for the food; selection was evidenced by the extra effort expended in finding the pork (and apparently bypassing chunks of beef).

This searching for resources is really the basis for the development of use—availability comparisons and explains why it originated with studies of diet. In most cases animals do not know the location of all foods in their home range, so dietary selection based on availability may be appropriate.

However, habitats are not spread around like chunks of pork and beef, but occur in large patches, the locations of which are known by the animals; thus habitats are probably more like McLellan's whole steer and whole pig than the cut up chunks of meat spread randomly around.

Consider some actual examples of how observed use and use versus availability can lead to disparate interpretations of selection. Prayurasiddhi investigated use and selection among two large ungulates, gaur (Bos gaurus) and banteng (B. javanicus), in Thailand. He differentiated two general study area boundaries, one of which more closely matched the area that his radiocollared animals used most intensively.

He also used actual home range boundaries as a third representation of the study area and hence the available habitat. He found that this variation in the area considered to be available habitat resulted in drastic differences in perceived habitat selection.

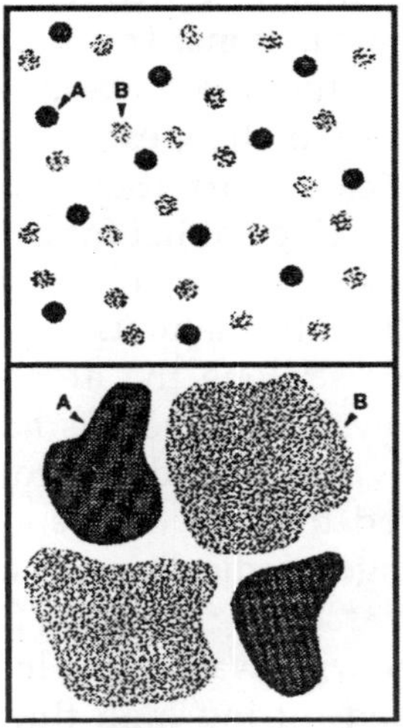

One habitat that received 46 percent of use by gaur was deemed to be selected for, whereas another habitat that received 45 percent of use was seemingly "selected against." In one case banteng were judged to be unselective in their use of a habitat in which they spent 75 percent of their time.

Prayurasiddhi recognized these difficulties and decided to evaluate seasonal changes in habitat selection and habitat-related differences among species based on use alone, rather than use compared to availability.

In another analogous situation, Macdonald and Courtenay found that crab-eating zorros (foxes, Cerdocyon thous) in Amazonian Brazil spent most of their time (64 percent) in wooded savanna and scrub habitats; however, because these two habitats were abundant within the home ranges of the study animals, they ranked the lowest in apparent preference (sixth and seventh among seven defined types) based on a comparison of use to availability. During the wet season, however, when lowland habitats flooded, the zorros used upland wooded savannas and scrub habitats even more, making the apparent preference.

Ranking for these rise; in reality, the area of available (not flooded) lowland habitats diminished, but this reduction could not be measured and therefore was not taken into account in the use:availability calculations. The authors realized that the seasonal difference in apparent habitat preferences was thus an artifact of unmeasured changes in availability.

In another such case, Garrett et al. found that tidal flats were the principal foraging habitat for bald eagles (Haliaeetus leucocephalus) and observed that nearly one-fourth of their perch sites were within this habitat.

That is, based on use alone, this area was clearly attractive to these birds. However, because this habitat was so widely available, especially at low tide, a comparison of use to availability suggested that the eagles avoided it. Apparent preference thus changed radically with tidal fluctuations.

A problem with assessing selection from use alone is that use often does change with availability. In the eagle study, tidal flats were rarely used during high tide because they were largely unavailable. The eagles may have preferred tidal flats to other habitats, but use of these areas depended on availability, up to a point.

When availability exceeded some threshold they did not continue to increase their use, and when it declined below some threshold their use of the area ceased. Between these thresholds some relationship may exist between use and availability. However, relationships between use and availability probably vary among habitat types, and within each habitat may depend on what other habitats are available.

It might seem that site attribute studies avert these problems because selection is inferred not from differences between use and availability, but from differences between used sites and other sites. However, the comparative sites are taken proportionately from those available, so the comparison is still, in a sense, the same as in the use—availability design.

Consider a study that locates 10 nest sites in each of two available habitat types and then compares the characteristics of these to 20 randomly chosen points. If one habitat type is much more common than the other, then the 20 randomly chosen points fall mostly within the more common habitat; this sampling could result in possible differences between the average characteristics of the randomly chosen sites and the real nests, suggesting that the species selected for nest sites in the less common habitat type.

This logic falters, though, for the reason highlighted in McLellan's example of the buffet, in which a person chose equal portions of beef and pork. In that example, the equal choice of the two types of food was clear evidence of nonselection, or equal preference, regardless of availability.

Accordingly, in the hypothetical case of the 20 nests, one should question why, if there was strong selection for one habitat type, did an equal number nest in each type? The conclusion that this represents evidence of preference for the less common type is (just as in use—availability comparisons) based on the assumption that if individuals had no preference, they would chose sites.

Randomly, which is typically thought to mean in proportion to availability. However, suppose each individual, having no preference, just flipped a coin, so to speak, to decide which of the two types of habitats to nest in each year. The result would be as posed in the example, an equal number nesting in each type.

There are two other problems (really two aspects of the same problem) with detecting selection using a site attribute design. James and McCulloch showed that a species could be highly selective for a certain habitat component, as indicated by a low variance among values of this component at the sites it chose, but if the mean value for the component at selected sites approximately equaled the mean in the available habitat (where the variance was much higher than at selected sites), then selection might not be detected (type II error).

Conversely, Rexstad et al. tested multivariate statistical procedures that are commonly used in site attribute studies and found that significant results, which in a real study would be indicative of selection for certain habitat components, were often derived from a collection of meaningless data (type I error). Taylor attributed Rexstad et al.'s spurious.

Results to ill-defined hypotheses and inappropriate use of statistical procedures; however, Rexstad et al. showed that their study mimicked the majority of published applications of these techniques. North and Reynolds expanded on Rexstad et al.'s concerns that the statistical procedures used in site attribute studies may yield misleading results because assumptions are commonly violated.

INFERRING HABITAT QUALITY[H]

If habitat selection studies were simply an attempt to better understand behaviour and natural history, errors would merely be a setback in scientific inquiry. However, these studies are typically designed to provide management guidelines.

Habitats are managed based on supposed importance in terms of fitness for members of the target population. High-quality habitats, by definition, produce animals with high reproduction and survival, and hence a population with a high growth rate (and high yield, for harvested species).

However, the assumption that we can infer habitat quality or suitability from studies of habitat selection—that selection, even if accurately measured, is directly related to each habitat's potential contribution to individual fitness and hence the population's growth rate—represents what the second fatal flaw in the process of habitat evaluation.

One of the best examples of a purported relationship between habitat selection and fitness was provided by a study of leaf-galling aphids (Pemphigus betae) that parasitize narrowleaf cottonwood trees (Populus angustifolia). These aphids overwinter in tree bark, then over a short period in the spring migrate up the tree and become entombed in expanding leaf tissue, where they reproduce.

Whitham found that aphids apparently selected large leaves over small leaves, even though large leaves were less available (less than 2 percent of leaves were over 15 cm; more than 30 percent were 5 cm or less); moreover, those that colonized large leaves had better survival and reproduction than those colonizing small leaves.

From this evidence it appeared that aphids had the ability to select leaves that offered the highest fitness; if not, the size distribution of colonized leaves would have matched the size distribution of leaves available on the tree (i.e., use would have equaled availability). However, Rhomberg observed that large leaves tended to be those near the tips of twigs (where they receive more light), and these opened a few days later than basal leaves.

When the aphids begin their migration, basal leaves are already too mature to enable them to form a gall, so they must colonize the more distal leaves; thus, the basal leaves are in essence nonhabitat. The migration appears to be timed so that the aphids inevitably colonize leaves that are destined to be favorable to their fitness, but individually, they do not select these among less favorable habitats.

This is not to say that other animals do not select habitats favorable to their fitness; certainly they do. But if the comprehensive and seemingly compelling data on a simple system such as aphids on cottonwood trees can be misinterpreted, then clearly there is much room for misinterpreting relationships between habitat selection and fitness in more complicated systems.

The presumed link between selection and fitness is rarely tested, partly because such tests are difficult, but also because this relationship is viewed more often as fact than as assumption. However, there are many reasons why this relationship may not hold. Habitats that are used infrequently may be more important than suggested by the time spent there.

Conversely, habitats used preferentially for activities that require a lot of time (e.g., resting) may be less important than indicated by their use; a resting habitat may be substitutable, with little effect on fitness, whereas a specific foraging habitat (or a water hole in an arid environment) may be more essential, even if not used very much. Differentiating activity-specific habitat use may help alleviate this difficulty, although this, too, is seldom done, and moreover, does not necessarily circumvent the problem.

For example, Powell studied winter foraging of fishers, which preyed heavily on porcupines (Erethizon dorsatum) but spent a disproportionately small amount of time hunting in upland hardwood habitats where porcupines were common. The reason, he found, was that fishers rapidly located porcupines at known den sites, thus minimizing their search and chase times. In contrast, fishers had a harder time hunting snowshoe hares, so they spent a larger amount of time in lowland conifer habitats where hares were more common.

Clearly, the relative importance of the two habitats to fishers was not reflected by their time spent hunting in each. For animals that feed on a variety of different foods, a mixture of different habitats may be more beneficial in many respects than a single, highly preferred type, and the time spent in each may be a poor indicator of importance of either the specific type or the overall mix. Omnivorous Egyptian mongooses (Herpestes ichneumon), for example, favour one specific habitat for resting, but use a mix of habitats when feeding.

Fitness also may be affected by predation and interspecific or intraspecific competition. These factors can change the cost:benefit ratio of a habitat and hence alter an animal's habitat use. Cowlishaw found that baboons (Papio cynocephalus) spent more time feeding in a habitat with poor food but a low risk of predation than in a food-rich habitat with a high risk of predation. Fitness was probably enhanced by this choice of habitats.

Fitness certainly was enhanced for a herd of caribou in Ontario, Canada, that spent most of the time on an island. The nearby mainland had higherquality forage but also a high density of wolves. Ferguson et al. found that the island occupants sacrificed nutrition, which was reflected in smaller body and antler size as well as starvation of their calves, for overall higher survival, which enabled this population to persist while herds that remained on the mainland perished from predation. One could argue that the animals on the mainland made poor choices in habitat selection.

Animals may not always correctly perceive risks, especially in a novel or changing situation, so their choice of habitats may not necessarily be best in terms of their fitness. Pollution is one such risk that is rarely evaluated in habitat studies. Consider Mallory et al.'s study of the effects of acid rain on habitat quality for common goldeneyes (Bucephala clangula).

Acid rain killed fish in lakes, which enabled proliferation of invertebrates, thus providing more food and enhanced reproduction for goldeneyes. However, these ducks may inevitably suffer other consequences of the acidified habitat. In some situations the choices made by animals may maximize fitness over the long term (e.g., if the changes are ephemeral); in other cases animals may require time to adapt.

Suppose that an animal uses alternate habitats more often when the normally preferred habitat becomes less profitable because of increased density and hence competition for resources. If the alternative habitats are more available than the preferred habitat, then use of these alternative habitats can end up higher than that of the otherwise preferred habitat. Situations like this seemingly justify considering habitat use in terms of availability.

However, Hobbs and Hanley showed both intuitively and with computer modeling that intraspecific competition may alter use: availability ratios such that they may not reflect the true quality of a habitat in terms of its potential contribution to population growth. Certainly interspecific competition and predation are integral components of habitat quality, and managers concerned about a given species should not evaluate habitats independent of these factors. However, perceptions of habitat quality that are affected by density of the targeted species are an artifact of the method.

Recall, as Peek noted, that habitat preference should be innate, and intricately tied to fitness; if our perception of preference declines as the preferred habitat becomes more densely populated, then our perception is clearly mistaken. Hobbs and Hanley concluded that "use/availability data inherently reflect differences in animal density among habitats" (independent of social behaviour) and therefore reveal little about habitat quality. Consequently, they recommended abandonment of the use—availability design in favour of an approach that enables a direct investigation of the link between habitats and fitness.

Site attribute studies often focus on sites of biological importance (as reviewed earlier) and therefore may provide more direct insights into habitat variables that affect fitness. However, as pointed out by North and Reynolds, conclusions from studies of this sort are based on two fundamental, but questionable, underlying assumptions: that used sites are in suitable habitat and that unused sites are in unsuitable habitat.

Consider the study by Meyer et al., who observed significant habitat-related differences among sites that were occupied versus unoccupied by spotted owls (Strix occidentalis). They found, as expected, that these owls tended to select sites having a large proportion and large patches of old-growth forest, but suggested that the owls could tolerate certain amounts of clearcut and young forest (based on the fact that used sites contained such areas).

Their reproductive data were inconclusive about what the owls could tolerate from a population standpoint, but Meyer et al nonetheless asserted "that resource selection probability functions are a reliable tool for assessing capability of landscapes to support northern spotted owls." Hence they concluded that their habitat selection model "can be used to predict the probability that a given landscape mosaic will be a suitable spotted owl site" (emphasis mine).

To avoid assumptions related to suitability in site attribute studies, North and Reynolds proposed that instead of comparing habitat characteristics of used sites with unused or random sites, used sites be categorized by their intensity of use (e.g., 1–2 percent of locations = low use, 3–10 percent medium use, more than 10 percent = high use), and habitat characteristics compared among such categories (i.e., excluding unused sites) with a form of logistic regression. Thus this is not really a site attribute design (according to my definition), nor is it even applicable to the types of situations in which site attribute studies are commonly used (i.e., in which an animal remains at a single site for a fairly long period of time).

Rather, this procedure represents just another means of investigating habitat characteristics associated with differing degrees of use. Whereas these methods may avoid many of the statistical and general assumptions of site attribute studies, they digress further from the issue of habitat-related fitness because the studied sites, instead of being of some special biological significance, are simply points where the animal was located some number of times. As North and Reynolds recognized, intensity of use may not be a good indicator of habitat quality.

ADVANTAGES AND PROBLEMS OF THE DEMOGRAPHIC RESPONSE DESIGN

The best measure of habitat quality would be a test of its effects on demographic parameters such as population growth and carrying capacity. Such tests are extraordinarily difficult in most situations, as evidenced by the scant published studies of this nature. Because habitat-specific density is actually a reflection of differential habitat use, investigations of habitat-related density suffer the same drawbacks as studies of habitat use.

Density tends to be an ineffective measure of habitat quality because it may fluctuate widely, is subject to sizable errors in estimation, and may be largely influenced by social factors. Van Horne gave several examples of situations in which juveniles were restricted from settling in the best habitats and thus accumulated in large numbers in poorer quality habitats.

She indicated that such circumstances are likely to be common among generalist species with high reproductive rates and a social hierarchy. For these species in particular, then, habitat-specific density would probably be a poor indicator of habitat quality unless the population is well below carrying capacity. A good example was provided by Messier et al., who showed that density of muskrats (Ondatra zibethicus) during a general population increase swelled 30- to 90fold in low-quality habitats but much less in high-quality habitats.

Considering that, in general, animals in poor-quality habitats should be trying to leave and ones in high-quality habitats trying to stay (and keep competitors out), Winker et al. posited that turnover rate would be a better index of habitat quality than density. They measured turnover rates for wood thrushes (Catharus mustelinus) by examining recapture rates and telemetry movement data; low-quality habitat was defined as that in which recapture rates were low and many radiotagged birds were transient visitors.

They found density and habitat quality, assessed in terms of turnover rates, to be inversely related. Notably, Winker et al.'s turnover rate model may not be applicable to other species, even other territorial, noncolonial songbirds, some of which are preferentially attracted to habitats occupied by conspecifics, which they use as a cue to habitat quality. Conspecific attraction tends to perpetuate use of the same areas across generations, so even if habitat quality deteriorates, high densities may be maintained through tradition.

Other competing species or unidentified confounding variables also may weaken the linkage between habitat quality and density. Maurer measured density and various habitat characteristics for five species of grassland birds; the habitat models developed to explain species-specific density in one study area were inexplicably poor predictors of density in a nearby area with similar habitat.

Kellner et al. found that density was positively related to reproductive success in only 7 of 17 bird studies that they reviewed; more studies showed a negative relationship. Sherry and Holmes felt that density should be relied on as an indicator of habitat quality only if it is corroborated by other data, as was the case in their study, in which population density and weight loss of wintering migrant birds were correlated (high weight loss in areas with low densities) and both were related to habitat type.

Reproduction and survival data may be more apt to reflect real influences of habitat on demographics. However, reproduction and survival are probably also tied to habitat in a complex manner. For example, a number of studies observed a direct relationship between cover and the survival (and thus density) of voles (Microtus spp.), but a lower threshold exists below which reductions in cover have little effect on vole density; above the threshold, survival and density increase but eventually reach an upper asymptote.

The vole studies found that cover provides food as well as protection from predators, and also may affect microclimate, activity patterns, and interactions among conspecifics, all of which affect the cover—density relationship. Moreover, male and female voles have different responses to varying cover, and cover—demographic relationships tend to be different for other small grassland rodents.

Each of the various habitat components that relate to an animal's fitness probably has thresholds, asymptotes, and inflection points, and these limits may vary with the mix, shape, size, and juxtaposition of habitat components available; however, few attempts have been made to assess any of these factors individually, let alone in combination.

Several studies also found that density-dependent effects may reduce reproduction or survival independent of habitat quality or may even result in higher fitness in low-quality, less crowded habitats. Zimmerman found that nesting success of dickcissels (Spiza americana) did not differ between habitats and was unrelated to density, but females nested preferentially in the habitat preferred by males; males chose this more heterogeneous habitat because they could sequester more nest sites and thus mate with more females.

In other studies, reproduction and survival were found to be unrelated to measured habitat variables, despite evidence of habitat selection, possibly because of confounding effects of weather, human disturbance, measurement error, and other factors. Even experimental demographic response studies have been plagued with unexpected variations in confounding variables.

Nevertheless, some carefully designed studies observed links between habitat and reproduction or survival and through further investigation discovered the underlying causes.

A final major problem, discussed previously in relation to use—availability and site attribute designs, is scale. Levin showed clearly that there is no single correct scale for studying ecological relationships. Animals view and react to their environment at various scales.

Human perceptions of ecological systems are inescapably biased or incomplete because they are filtered by the observational scale chosen for the investigation. Some demographic response studies have recognized this and have adopted a multiscaled approach. For example, Orians and Wittenberger found that densities of yellowheaded blackbirds (Xanthocephalus xanthocephalus) were higher on marshes with higher food (insect) abundance but that density of blackbird territories within these marshes was related to vegetational structure, not food.

The authors postulated that these birds may not select food-rich territories because they often hunt outside their territories. Also, they establish their territories before the full emergence of insects and hence may not be able to predict future food abundance on a scale smaller than the marsh level.

Pedlar et al. developed habitat models on two different scales to explain variation in raccoon (Procyon lotor) density. One model was fit to macrohabitat features and the other to microhabitat variables, after which the two were combined to form a more comprehensive model relating density to habitat at both scales.

Morris observed both macrohabitat and microhabitat differences among several species of small mammals, suggesting habitat selection on both scales, but found that variation in density was much more evident at the macrohabitat level. On a larger scale, Dooley and Bowers discovered, counter to their expectations, that densities and population growth rates of voles were higher in patches within a fragmented landscape than in an unfragmented landscape, whereas total population size was higher in the unfragmented landscape (because more total habitat was available). Landscape fragmentation caused overall habitat loss but, on a finer scale, enhanced reproduction within individual habitat fragments.

Similarly, Brown and Litvaitis noted that some mammalian predators that hunt preferentially in forests nonetheless often exist at lower densities in homogeneous forests than in forests interspersed with disturbed areas. Each of these examples represents cases in which a single-scale investigation would have failed to detect habitat variables affecting population demography.

The previous examples all concerned spatial scale. Time scale may be equally important. The demographic value of a habitat may become evident only in the long term, after a population has been subjected to the stresses of a periodic drought, severe winter, or failed food crop.

APPLICATIONS AND RECOMMENDATIONS

A great deal of effort continues to be invested in habitat-related studies of wildlife. In the United States, federal land management agencies in particular have focused on developing formalized procedures for evaluating habitat for wildlife. The procedures that have been adopted rely on models derived by species experts, who in constructing these models tend to rely more on experience than on empirical data. Therefore, the models are really hypotheses in need of testing.

However, because these models hypothesize explicit relationships between habitat attributes and animal populations (so-called habitat suitability indices), they cannot be rejected or accepted in normal scientific fashion; that is, none of the relationships are likely to be exactly correct. Thus it seems inappropriate to suggest, as is common parlance, that they should (or even could) be "verified" or "validated."

Several attempts have been made to test such models. Often this process is circular, involving just another panel of experts making qualitative assessments. In other cases, models have been tested using results of a study on habitat use or use relative to availability, with the inevitable associated shortcomings. In some instances, habitat management prescriptions based on "common knowledge" or expert opinion have, through collection of better data, been proven faulty. One case in which model-derived habitat scores for individual home ranges were compared with reproduction, juvenile growth rates, and home range size, but no significant relationships were observed.

Often, models have been tested by comparing habitat-specific densities to model predictions. However, even if a model explains a significant portion of the variation in density, the data collected to test (or purportedly validate) the model could be better used to modify it or build new one. In most cases, habitat models have proved to be poor predictors of animal density, indicating either defectiveness of the model, lack of a clear habitat—density relationship, or effects of other confounding factors, such as hunting pressure, which are also habitat-related.

Bender et al. found that if the variance around the estimated values of the model inputs were taken into account (a process that is not commonly done), suitability scores for a variety of habitats that appeared very disparate were not significantly different; that is, the parameter estimates were not precise enough to even enable the model to be tested.

A major inherent but generally unstated (maybe unrecognized) assumption of habitat suitability models is that high-quality habitats (i.e., habitats that confer high fitness) are in fact suitable. Explicit tests of this assumption are rarely conducted, yet counterexamples exist.

In one case, a habitat suitability model for Florida scrub-jays (Aphelocoma coerulescens) correlated well with demographic performance, but most of the area was found to be a population sink where mortality exceeded reproduction. Apparently, as birds competed to occupy the best habitats, they were less alert to predators, and thus suffered high mortality. In another example,

Kirsch found that interior least terns (Sterna antillarum, least terns nesting in noncoastal areas) selected high-quality nesting habitats, but possibly because of disturbance, their productivity was not sufficient to maintain population size (i.e., the nesting habitats were unsuitable).

Lomolino and Channell observed that remnant populations of endangered mammals often occur near the periphery of their former ranges; because the periphery of the range represents the edge of suitable habitat, studies of habitat suitability of endangered mammals based on habitat use in existing populations are likely to be misleading. These are not gratifying results.

It has not been for lack of effort, expense, or analytical developments that relationships between habitats and population growth often elude detection. It is simply the complexity of the interactions between animals and their environment that make such relationships deceptively difficult to understand.

A case in point is the spotted owl, which has undergone intense scrutiny because of its threatened status and apparent proclivity for old-growth forest in a region where the economy is tied largely to timber; despite a plethora of habitat studies, significant debate persists among ecologists as to the critical habitat requirements of this species, and why it prefers old-growth forest.

However, substantive flaws in the most commonly used techniques for studying wildlife—habitat relationships apparently have not been widely recognized. It is time to reconsider the ways these techniques are used in evaluating habitat quality.

Studies of the use of habitat have merit, but also many limitations. Habitats in which an animal spends a large proportion of its time are clearly selected among others available. Even if widely available, frequently used habitats are certainly not "selected against" or "avoided" (except maybe among life forms that have no memory of where they were). However, frequent use suggests nothing about habitat importance or substitutability in terms of fitness. Correspondingly, infrequent use may not be indicative of lack of suitability.

A habitat may be used infrequently because it serves little value, because its value can be extracted in a short amount of time, because it is not readily available, or because access is constrained by threats (social pressures, competition, predation) or physical barriers. A high use:availability ratio might suggest (correctly) that such a habitat is more important than indicated by its infrequent use.

Studies of habitat use would benefit greatly from replication. Significant insights might be gained from comparisons of habitat use and use:availability among individuals, among groups of individuals in different portions of a study area, among study areas, among time periods, and so on. If individuals are disparate in their use or apparent selection of habitats, inferences regarding habitat quality become more equivocal.

In contrast, if the data are partitioned and the subsets show consistent patterns (e.g., use:availability ratios for each habitat are similar among individuals despite large differences in availability within individual home ranges) or if curvilinear relationships between use and availability can be ascertained, inferences are strengthened; even so, studies of this sort provide only a superficial understanding of the effects of habitat on population dynamics. Certainly no strong prescriptions for habitat manipulation are warranted from interpretations of selection based solely on observed patterns of habitat use.

Site attribute studies tend to provide stronger inferences about habitat selection. However, in identifying myriad habitat characteristics that are apparently preferred, even for activities that impinge on the animal's survival and reproduction, these studies still cannot assume that population growth would be significantly higher in more "ideal" settings.

Controlled experiments can help sort out the factors important in the animal's selection processes and can thus provide a better understanding of its behaviour, but without corresponding demographic measurements, the importance of various habitat components in terms of their contribution to the animal's fitness cannot be appraised. Kirsch:

"Unfortunately, proximate habitat features may not indicate habitat suitability, nor do they reveal the possible selective pressures that influence habitat selection in a system. One must measure components of fitness, determine factors that influence fitness, and relate fitness and factors influencing fitness to habitats or habitat features."

Demographic response studies are the only means of truly evaluating the relative importance and suitability of habitats for supporting animal populations.

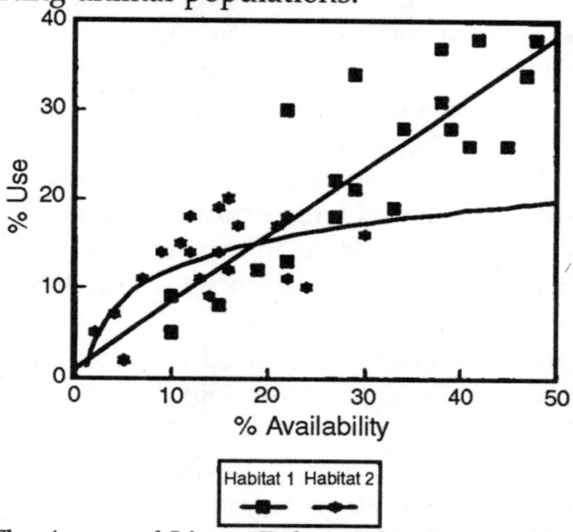

Fig. The Assumed Linear Relationship Between use and Availability of Habitats can be Tested, to an Extent, by Plotting use and Area of each Habitat in Individual Home Ranges.

The figure depicts such a plot for 20 hypothetical home ranges (each point represents habitat use and availability in one home range) and two (of several) types of habitat (sums for both use and availability for all habitat types for each individual would total 100%).

The data suggest that increased availability of habitat 1 prompted a corresponding increase in use. In habitat 2, however, the animals showed an asymptotic relationship between use and availability; 10% availability seemed to represent an approximate threshold, above which use no longer increased. That is, if at least 10% of their home range was composed of habitat 2, the animals could obtain whatever resources they needed from this habitat by spending 10–20% of their time there. A similar threshold may exist for habitat 1 at availabilities > 50%; assumptions should not be made beyond the data.

These should be given greater emphasis. Oddly, Hansson suggested that habitat-specific survival might confuse perceptions of habitat selection. In his small mammal study, perceived habitat selection appeared to be a consequence rather than a cause of differential survival. It seems to me that such knowledge of habitat-specific survival is exactly the desired objective; habitat selection studies are just an indirect approach toward this end.

For many species, habitat-specific densities may be easier to measure than habitat-specific reproduction or survival, but density studies may yield uncertain or misleading results because density is the end result of various processes, both demographic and behavioral, each with potentially different habitat-specific responses.

Controlled experiments should be used more often in assessing effects of habitat on demographic parameters such as density or reproduction; however, even elegant experiments may produce unexpected results in complex natural systems.

Where experimentation is unfeasible, comparisons of reproduction and survival can be made among individuals or groups of individuals with different patterns of habitat use. Partridge warned that such comparisons may be confounded by competition among individuals:

Those living in preferred habitats may be dominant individuals that are naturally fitter. That is, they live where they do because they are fitter; they are not fitter because of where they live. This is an important consideration, but it is not applicable to all situations; it depends on the social structure and population size relative to carrying capacity. One way to circumvent the problem is to compare population parameters across multiple study sites with differing habitat compositions.

Often, though, it may be appropriate to assume that individual physical differences are not the cause of differences in habitat use, or even if they are, that the most fit individuals are choosing the best habitats, thereby justifying comparisons within a single study site.

For studies in a single study site, an effective design would be to monitor habitat use of various individuals (e.g., using radiotelemetry), and then compare their frequency of use of different habitats or the habitat composition of their home ranges to their eventual reproduction and survival. Many studies of birds have used this sort of approach to examine relationships between habitat characteristics of nesting sites and nesting success.

However, few attempts have been made to relate reproduction or survival to variation in habitat use within home ranges. In one example, Aanes and Andersen observed a relationship between survival of roe deer (Capreolus capreolus) fawns and the habitat types that they used. Similarly, Aebischer et al. found that habitat use by individual pheasants was related to their survival.

In another example, Fuller observed that wolf (Canis lupus) territories were limited to areas with low road density (a habitat component that relates to human access) and then looked for (but did not find) a relationship between survival of radiocollared wolves and road density within individual territories.

In another study of the same species, Massolo and Meriggi found relationships between reproduction (documented presence of pups) and various natural and anthropic habitat features, including road density. In studies in which data on survival and reproduction are not obtainable or too scant for making comparisons with habitat use, surrogates (e.g., body size, home range size) may, in some cases, be appropriate and could add measurably to assessments of habitat quality.

Surprisingly, even intensive, long-term studies of radiocollared animals with detailed data on both habitat use and demographics have typically not attempted to relate the two. This should be a goal for the future.

Chapter 4

Terrestrial Vertebrates

Why study food habits? Probably one of the most fundamental questions that ecologists attempt to answer is, "What resources does a particular species require to exist?" Indeed, the first principle among wildlife ecologists is to have a thorough understanding of the food, cover, and water requirements of an animal before initiating any effort to alter the factors that may be limiting it.

Information on food habits is therefore an important introduction to the natural history of any species. This has been a justification for many studies of food habits of vertebrates and is still a valid reason to investigate the diet of any species when little information is available. Food habits have been investigated for a variety of other reasons.

Such information is essential in understanding the potential competitive interactions among sympatric species or in determining how the foraging patterns of individuals affect community composition. For example, how does grazing by wildebeests (Connochaetes taurinus) affect the diversity of grasses and forbs? Does predation by lions (Panthera leo) limit that same wildebeest population?

A simple list of foods used by wildebeests or lions will not answer these questions. However, determining the biomass consumed and abundance of alternative forage or prey is an important first step in understanding how these two species influence community composition.

In human-dominated landscapes, information on the food habits of common terrestrial vertebrates has been useful in understanding the "economic food niche" of many species.

Losses of livestock, agricultural crops, or game populations are serious economic concerns. Limiting these losses is a major charge of government wildlife management agencies.

Historically, efforts to control depredating wildlife have included indiscriminate attempts to reduce populations of the offending species. Well-known examples of such an approach include the efforts to reduce large carnivores in North America. Detailed investigations have revealed that the actual nuisance individuals may be only a portion of the population and depredation may be restricted to a limited time period.

A recent motivation to examine animal food is the emergence of environmental assessments during large-scale habitat alterations, such as road construction and commercial timber harvesting. Biologists must identify the important food and cover resources in an affected area if they are to mitigate the effects of these activities.

CONVENTIONAL APPROACHES AND THEIR LIMITATIONS

DIRECT OBSERVATION

Direct observations have been widely applied to document the forage or prey used by a variety of species. Individual animals or groups are observed through binoculars as they graze or feed on an animal carcass. Observations at bird nests also have provided information on foods brought to juveniles. The basic approach is simple and relies on limited equipment. For researchers studying herbivores, bite counts or feeding minutes by plant species are recorded.

These values can then be translated into relative occurrence in the diet by comparing total bites or minutes of foraging and the contribution of each species to the total observed. Biomass consumed can be approximated by estimating the average mass per bite for each species incorporated in the diet. Additionally, direct observations are useful in identifying differences in foraging among sexes or age classes. Unfortunately, this technique is hampered by several limitations.

Observations are usually limited to species that occupy open habitat (grassland, savanna, or tundra) and forage during daylight periods. Consistently identifying the grasses and forbs consumed by herbivores may limit the application of this technique to habitats where forage diversity is limited or species are easily differentiated.

For wideranging animals, the observer must follow subject animals and thus may affect their foraging behaviour or prey availability. Among carnivores, observations are usually limited to identifying large prey that are not eaten whole.

LEAD ANIMALS

Concern about the suitability of direct observation has prompted some investigators to use tame, hand-reared animals to investigate food use. Subject animals are followed closely by the researcher, allowing accurate identification of the foods consumed and those avoided. Lead animals provide an opportunity to observe the consumption of items that are consumed whole.

(e.g., fruits and small forbs) or where there is little evidence of consumption. Although this approach may eliminate some of difficulties associated with direct observations, the diet summaries reported for lead animals have been criticized as being artificial. Physiological condition, hunger, presence of conspecifics, and previous foraging experience may affect food use by tame animals and result in erroneous conclusions. Although this approach has had only limited.

Application among carnivores, it may be useful under unique circumstances (such as a falconer using a trained raptor) to investigate differential vulnerability of prey if captured prey are compared to a random sample of the target population.

FEEDING SITE SURVEYS

Feeding site surveys are among the earliest approaches used to investigate food habits and were initially developed to determine the foods of livestock. This method relies on an inventory of plants consumed or identification of prey remains.

Survey of food Remains

Because of the limitations imposed by cover or nocturnal activity, the food habits of many species cannot be observed directly. This technique obviously depends on the ability of the investigator to locate sites where feeding has occurred.

Table. Evaluation of Methods Used to Investigate vertebrate food habits

Method	*Advantages*	*Disadvantages*
Direct observation	**Inexpensive Simple age or sex** differenes if these can be determined	**Limited to diurnal periods and open habits** Persence of observer in the field affects activity of consumer and potential prey
Lead animals	Provide very precise information selection can be investigated by sampling available food	Expensive limited size Results considered artifscial unless subject acclimated before data collection
feeding site surveysfood remains	Can prowide summary of major foods consumed	small or completely consamed foods are not sorveyed
	can cstimate biomass consumed	Cannot examine age or sex differences in food use
before—after	Can estimate biomass consumed	must be able to differentiate consumers
Exclosures	Useful in evaluating long term effects of herbivory on plant community dynamics	provides information only on major plants consumed No information on diets of age or sex categories Must be able to distinguish foraging

		by sympatric herbivores
Gastrointestinal tracts	Able to examine age and sex differences Can examine other parameters (e.g., physical condition and reproductive rates) from other carcass samples	Samples usually limited to legal harvest by hunters or rappers

If more than one species is known to forage in the area, than there must be a way to differentiate consumers (e.g., foraging by lagomorphs versus foraging by ungulates). Among herbivores, grazed grasses and forbs or browsed twigs are counted and converted into percentages of the total forage consumed. For woody plants, it is possible to estimate the biomass consumed by measuring the diameter of the twig at the point it was clipped and using regression tables specific for each species.

Such surveys may be biased by "invisible consumption" when an entire plant is consumed and there is no residual indicating the occurrence of that plant. For carnivores, this approach has been used to identify the age, sex, and physical condition of large prey, but provides no information on prey that are eaten whole or cached. A major disadvantage of this approach is the lack of opportunity to investigate sex or age differences in food use.

DIFFERENCE COMPARISONS

This approach is restricted to herbivores and relies on comparisons between used and unused plots or sites surveyed before and after foragers have passed through an area. Species consumed (based on inventory) or biomass removed can be estimated by clipping used and unused plots. An obvious limitation of used and unused sites is the ability of the investigator to locate comparable sites.

Also, unless the investigator is able to verify nearly exclusive use of a site by the herbivore under investigation, before and after comparisons may be clouded by a variety of herbivores foraging at the same site. In general, difference methods do not detect small differences in use. Therefore, they should not be considered unless use is expected to be more than 50 percent on the forage species.

A more accurate but labour-intensive technique for assessing use has been applied in the western United States. Stems of potential browse species are tagged and total length of annual twig growth is measured during autumn. In spring, the investigator returns and measures the length of annual growth that remains.

Although the technique does not directly measure biomass removed, the length of the twig removed is highly correlated with the amount of forage removed. A limitation of this approach is that it requires some knowledge of the plants that are likely to be consumed.

EXCLOSURES

A slight variation on difference comparisons is the comparison of used sites and sites where access has been restricted by an exclosure. Exclosures are usually wire fences or cages that limit access by herbivores but do not hinder plant recruitment or growth. Exclosures can reveal information on general food habits of herbivores when based on short-term differences between paired (fenced and open) plots.

This technique should not be applied to estimates of use during the growing season because protected plants grow at different rates than grazed plants. Probably the more common use of exclosures is to reveal the effects of herbivory on plant community composition. Here, investigators usually erect a smaller number of large exclosures that are monitored for 1–2 years or longer.

Periodic inventories are conducted to compare species composition and growth rates between exclosures and open sites. Exclosure studies have revealed that herbivores can affect the abundance, biomass, and diversity of plants. For example, moose (Alces alces) in boreal forests affect the recruitment of forage and nonforage species.

Long-term exclosure studies also have revealed that this herbivore not only alters the species composition of forest stands, but also can directly affect the structure of herb, shrub, and canopy layers. Although exclosures can help identify major forage plants and responses in plant communities, they are not an appropriate method to gain information on infrequently used forage.

POSTINGESTION SAMPLES

The most common technique for analyzing food habits of terrestrial vertebrates involves sampling either during or after the digestive process. Samples may be collected from various stages of digestion for use in identifying food habits of herbivores and carnivores. All postmastication sampling requires identification of materials that may not be easily recognized.

Feces or Pellets

Examining the content of feces or regurgitated pellets has become widely used because this approach is nondestructive and large samples can be collected. This technique has been applied with equal success to waterfowl, upland birds, and raptors as well as small and large mammals. To distinguish fecal samples, field guides are available that rely on size, shape, and colour to identify the source.

If a less ambiguous approach is needed, especially where sympatric species have feces or pellets that are similar, fecal pH or bile acid differences can be used to distinguish samples. A newer approach uses information on mitochondrial DNA from epithelial cells shed from the intestines of the animal that defecated or regurgitated the pellet.

Among herbivores, partially digested seeds and fruits can be identified macroscopically. However, most of the analysis of herbivorous materials relies on microhistological techniques to identify characteristic cells and structures of foods consumed. A reference collection of potential food items is crucial. Microscope slides must be prepared from all potential food plants in the same manner as sample materials. In addition, a collection of local seeds and fruits should also be made for reference.

Generally, less than half of what appears on a typical slide will be identifiable plant fragments. The cellular characteristics used to identify plant fragments are those that survive the mastication and digestive process and are generally composed of epidermal tissue. These include cuticle, stomata, cell walls, aperites, glands, trichomes, silica cells, druses, crystals, starch grains, and silica—suberose couples as well as general cellular configurations, size, and other structural characteristics.

Because the ratio of identifiable to nonidentifiable fragments changes during digestion and sample preparation and because certain browse species have a low proportion of epidermal material in relation to their biomass, correction factors may be developed to improve the approximation of diet composition.

Several researchers have recommended that hand-mixed diets be used to test the assumption that the actual diet matches the diet estimated from microhistological analysis. Others have suggested that the differences are too small to justify developing correction factors or that correction factors do not consistently improve the estimation of diet composition, particularly when the diets contain a variety of grasses, forbs, and browse.

Investigators of herbivore diets that relied on fecal analysis have been criticized for identifying fewer species than can be found in rumen samples. Generally, easily digested forbs are underestimated and the less digestible items are overestimated. Inaccuracy of the technique, particularly when applied to diets of mixed (grasses, forbs, and browse combined) feeders, has led some researchers to question its usefulness for herbivores other than grazers. Lack of experience and training in identifying plant fragments are often cited as the most important sources of error in using the microhistological technique.

In contrast to herbivore samples, fecal material and pellets from carnivores require minimal preparation. Depending on the size of prey and the level of mastication, much of the material in a sample from a carnivore may be identified using field guides to birds, mammals, and insects for comparison.

Other items may be identified through bones, teeth, hair, feathers, or scale patterns. Therefore, reference materials may include complete skeletons of vertebrates, samples of hair, feathers, scales of fish and reptiles, and exoskeletons of insects.

A collection of dorsal guard hairs of the mammals likely to be encountered is particularly useful because of the characteristic features of colour banding, medullary pigment patterns, and the morphology of cuticular scales.

Recently, several investigators have used postingestion samples to explore selectivity of prey consumption or differential vulnerability. These studies were based on skeletal remains in feces that could be distinguished to different sex or age categories.

Composition of prey remains was then compared to trapped samples. Although this approach has increased the information obtained using postingested samples, it is important to recognize that for larger prey it is not possible to determine whether the prey was killed or scavenged.

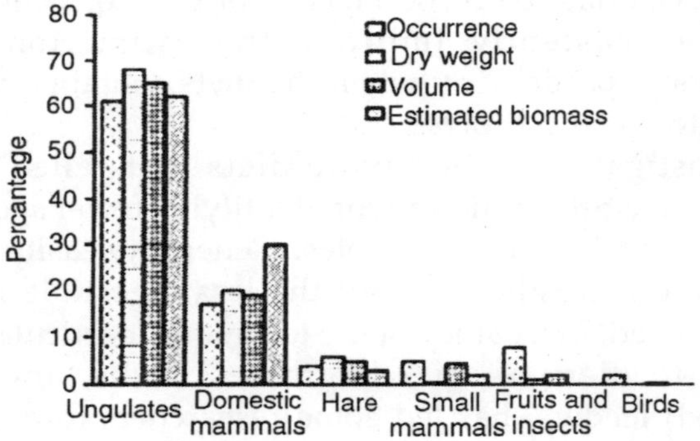

Fig. Comparison of four Methods used to Investigate Prey use by Wolves.

Probably the most contentious aspect of using postingested samples is the lack of an unambiguous method to quantify the contribution of specific foods to the total biomass consumed. A variety of approaches have been used, including frequency of occurrence of specific prey among samples, measured weights of remains recovered in fecal samples, relative volume of prey remains, and estimated biomass consumed calculated with digestibility coefficients. Evaluations of these approaches relative to ranking prey importance indicated that the most common inconsistencies occurred among small prey.

Fecal samples obtained from free-ranging wolves in Italy. Occurrence was the percentage of each prey relative to total occurrences of all prey. Dry weight was a percentage of all weight of all prey remains recovered in fecal samples. Volume was estimated using a reference grid and each prey was expressed as a percentage of the total volume of the feces.

Development of digestibility coefficients or conversion factors seems to have the greatest potential for estimating actual biomass consumed. Correction factors are generated by feeding known amounts of forage or prey, collecting the resulting fecal material, and weighing the contents. These results can then be used with volumetric estimates or weights of prey remains to create a diet profile. Ideally, an investigator should use consumer-specific correction factors. Hewitt and Robbins also recommended developing correction factors for specific foods.

Until such correction factors are derived, however, it may be useful for investigators to apply existing correction factors developed for general food or prey categories (e.g., foliage, fruits, nuts, small vertebrates) as long as the feeding trails used to generate the correction factors and their subsequent application are trophic-level specific (i.e., correction factors developed for carnivores are applied to carnivores and not to herbivores).

Although there may be interspecific differences in digestion efficiencies within a trophic level the biases associated with the subsequent estimates of biomass consumed will probably be smaller than those based on percentage occurrence or percentage volume.

GASTROINTESTINAL SAMPLES

Contents of alimentary tracts are generally collected only from wild animals with large populations because they usually involve sacrificing the animal. Investigators use this approach often on collections during legal hunting and trapping seasons. These harvests usually span only a short portion of the year, limiting the application of this technique.

For carnivores, harvest samples also may be biased by foods used as bait. On the other hand, information about the sex, age, and body condition of the sampled animal and volume of prey consumed is an advantage of this method.

Crop contents of granivores are a very different sample because only limited digestion has occurred. Often the investigator can separate and identify the majority of the sample. The volume of each seed type also can be estimated with a graduated cylinder or by the displacement of a known quantity of water in a burette. Seeds and fruit can be identified by comparison with a local reference collection or through the use of reference books.

Among herbivores, esophageal- and rumen-fistulated animals have been used instead of sacrificed animals. Fistulating involves installing a permanent device in the digestive tract of a living animal, allowing samples to be taken as food passes that point in the digestive process.

Fistulation has been used extensively in describing the diets of domestic animals, but rarely has this approach been applied to wild ruminants. Taming and hand-rearing fistulated animals are the only ways a researcher can approach them to collect samples. But as with using domesticated "wild" herbivores for bite count data collection, the time and money required are generally prohibitive and tame animals may not reflect the true food habits of their wild counterparts.

Emetics, flushing tubes, and manual expression of the gullet have been also been used, primarily on birds, to purge the upper portion of the digestive tract without harming the animal. More recently, nonlethal sampling of stomach contents has been successfully done on small vertebrates, especially reptiles.

EVALUATING THE IMPORTANCE OF SPECIFIC FOODS AND PREY

Use, selection, and preference have been applied

interchangeably when discussing food use patterns, resulting in some confusion. Use simply indicates consumption of a specific food. Selection implies that an animal is choosing among alternative foods that are available.

Use is selective if foods are consumed disproportionally to their availability in the environment. Preference is independent of availability. For example, animals can be provided different foods on an equal basis (cafeteria experiment) to determine preference among the foods provided.

AVAILABILITY VERSUS ABUNDANCE

Any evaluation of the selection of foods can be accomplished only with information on food availability. Unfortunately, the availability of forage or prey species can be difficult to estimate. Physical access to forage plants may be constrained by the reach of a herbivore or snow coverage. Snow and vegetation structure also may influence the availability or vulnerability of prey to carnivores.

Likewise, the presence of predators can affect the distribution of foraging activity by herbivores. Many investigators have relied on an estimate of food abundance as a surrogate to availability. Is this an appropriate compromise? Estimates of relative abundance are often in units that do not necessarily correlate with density or biomass, such as captures per 100 trapnights or individuals observed per kilometre of transect surveyed.

Obviously, large differences between consumption and relative abundance are pertinent to understanding forage or prey selection. However, studies that incorporate estimates of food abundance without considering the limitations of estimates of availability should be viewed as inferential.

CAFETERIA EXPERIMENTS

The ability of an investigator to provide equal access to all foods as a method of identifying preference is obviously limited to captive situations. Rather than simply identifying preferred foods, most researchers who have used cafeteria experiments have attempted to identify the components that affect diet composition and therefore complement field observations. For example, Klein observed that snowshoe hares (Lepus americanus) consumed twigs from sprout growth disproportionately less than twigs from older trees. Nutritional analysis (including protein content) did not explain this differential consumption.

Klein speculated that hares might have been responding to antiherbivore chemicals in juvenile plants. Bryant and colleagues later examined this relationship by extracting resins from juvenile plants and then compared consumption rates of resin-coated and uncoated twigs by captive hares, providing experimental evidence that these compounds do exist and may indeed explain foraging preferences by snowshoe hares.

INNOVATIONS

IMPROVEMENTS ON LEAD ANIMAL STUDIES

Recent improvements in the use of lead animals may cause this approach to become more commonly used if very detailed information on food use is needed. Hand-reared animals can be equipped with radiotransmitters and released into the wild several months before data collection. Lead animals are often visited by their handler to remain acclimated to humans, but should no longer be naive about natural foods. Use and availability can be accessed by walking with the animal as it feeds, recording use, and returning to measure availability along a marked trail.

USE OF ISOTOPE RATIOS

Biogeochemists have demonstrated the utility of comparing the relative concentration of various isotopes of carbon, nitrogen, and sulfur to reveal elemental cycles. This approach has been applied to examine current and historic food use patterns of some vertebrates. Essentially, the analysis of carbon isotopes can be used to determine the relative contributions of marine and terrestrial sources to an individual's carbon pool. In marine ecosystems, carbon enters as a bicarbonate and 13C is enriched relative to terrestrial ecosystems. This information can then be used to compare the importance of marine and terrestrial sources of forage.

For example, in the northwestern United States, diets of historic populations of brown bears (Ursus arctos) were assessed by examining isotopic signatures from hair and bone collagen from specimens collected 140 years ago. The relative abundance 13C also can be used to examine consumption of C 4 (many grasses) versus C 3 (trees and shrubs) plants. Nitrogen-15 is enriched at each step of the food chain, probably because the preferential excretion for lighter 14N in urine.

EXPERIMENTAL MANIPULATIONS

Although it is widely accepted that populations of vertebrates are ultimately limited by the abundance of food, there has been substantial controversy among ecologists on how food abundance affects population dynamics and community structure. This has led a number of investigators to experimentally manipulate food availability.

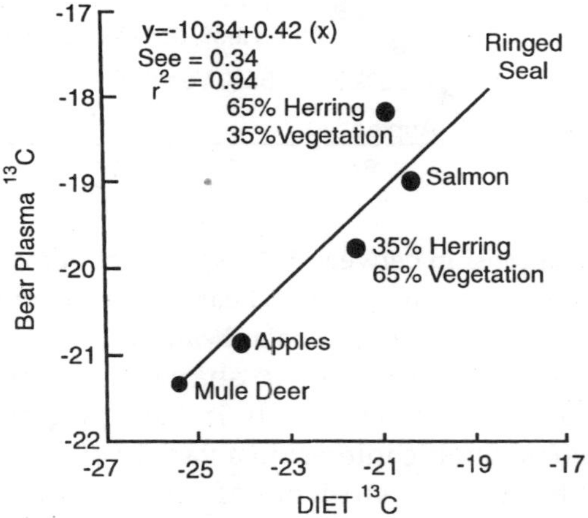

Fig. Relationship between 13C Signatures of the diet of Equilibrated Plasma in black Bears (Ursus americanus, (•) and Polar Bears.

Most of these studies were concerned with understanding how food supply affected such characteristics as reproduction and density, not with food selection or preference. In his review of more than 130 studies,

Boutin found that most studies increased food abundance. Among individuals with access to supplemental foods, there were some clear tendencies: reduced home ranges, increased body weights, and advanced age and timing of reproduction. Populations typically doubled to tripled in density where food was essentially unlimited.

However, food addition did not prevent declines in fluctuating populations.

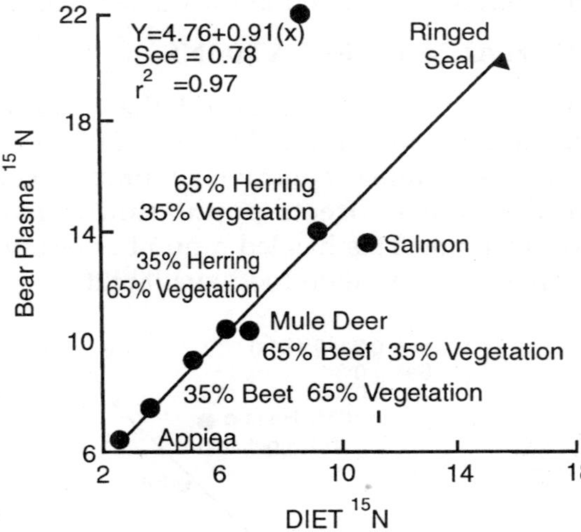

Fig. Relationship between 15N Signatures of the Diet of Equilibrated Plasma in black Bears (Ursus americanus, (•) and polar bears.

The ability of herbivores to shape the composition and structure of vegetation via their foraging patterns has generated considerable interest in a variety of ecosystems. This has also been the subject of several experimental manipulations.

For instance, Ostfeld and Canham planted seedlings and monitored herbivory by meadow voles (Microtus pennsylvanicus) to reveal that this small mammal can effectively delay the invasion of trees into old-field communities. Spatial variation in foraging by voles also may create a mosaic of woody and nonwoody vegetation in these sites.

THE ROLE OF FORAGING THEORY IN UNDERSTANDING FOOD HABITS

Most animals exploit a narrower range of food items than they are capable of consuming. In an effort to understand the constraints that may determine diet width, ecologists have organized their evaluations of food selection into a body of theory called optimal foraging theory. Two distinct approaches have developed to address this issue.

The first considers that an animal selects among various food or prey items that are distributed in some fashion (e.g., clumped) throughout a generally suitable habitat. The second approach examines how animals discriminate among various patches of habitat that vary in productivity and suitability and can be viewed as an evaluation of habitat selection.

Early efforts to understand diet width relied on evaluating potential food items in terms of cost (search and handling time) and benefit (energy). According to the basic assumptions of foraging theory, an animal should have a diet that maximizes energy intake and minimizes time to obtain nourishment (Schoener 1971).

As such, food or prey items are ranked by profitability and added to the diet as long as there is an increase in net energy intake. The optimal diet model provides several useful predictions. If handling times (the time needed to pursue, capture, and consume) are typically short, the consumer should be a generalist (use a wide range of foods or prey).

On the other hand, if handling times are long, the consumer should specialize on the most profitable foods. Consider prey selection by wolves (Canis lupus) that are usually in close proximity to large ungulates, such as moose. The time and energy required to capture a moose may be considerable. Wolves may specialize on the most profitable or vulnerable segments of the population (juveniles and older animals in poor condition).

Optimal foraging theory also predicts that a consumer

should have a broader diet in an unproductive environment or during lean periods than in a productive environment or periods of food abundance.

Although optimal foraging theory has provided an important platform for understanding consumer—prey relationships, the successful application of this theory to understanding diets of free-ranging vertebrates have been limited. The predictions of this theory are based on a series of assumptions that may not be justified.

The first is that the foraging behaviour exhibited by presentday animals was favored by natural selection and continues to enhance the fitness of animals; the second is that high fitness is achieved by a high rate of net energy intake. Numerous field investigations in the last decade have revealed that diet selection is probably the consequence of fairly complex interactions of external, internal, and phylogenetic factors.

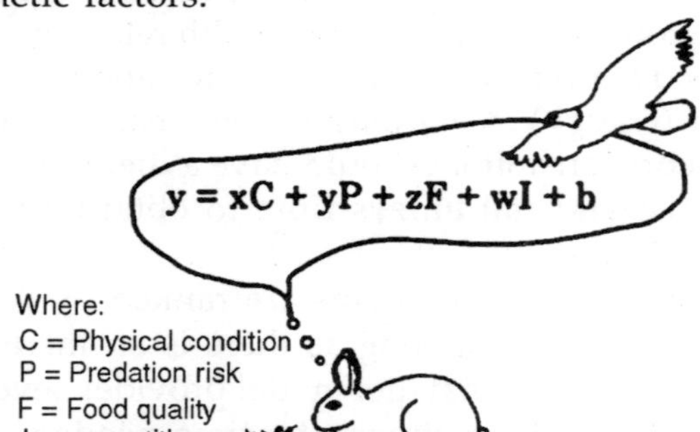

Fig. Internal and External Factors Affecting foraging Decisions by a lagomorph

External factors may include prey availability, risk of predation and social interactions (e.g., competition). Internal factors include animal condition or hunger, learned experiences, age, sex and reproductive state, macro- and micronutrient requirements, and concentration of toxins or distasteful compounds.

Phylogenetic factors include morphological constraints (e.g., mouth shape), sensory limitations, and physiological limitations. With such a complex array factors now known to affect foraging decisions, hindsight is quite clear: General models will probably fall short in contributing to our understanding of foraging patterns.

However, recent innovations (e.g., using phylogenetic comparative methods) and continued use of manipulative experiments will undoubtedly advance our ability to identify parameters that are influential in complex environments.

SAMPLE RESOLUTION AND INFORMATION OBTAINED

As we have seen, the many methods used to investigate vertebrate food habits do not provide comparable information. The approaches summarized in table differ according to the sample unit and resolution of information obtained. On one hand, exclosures provide information on relative use only by the population that is restricted by the fencing.

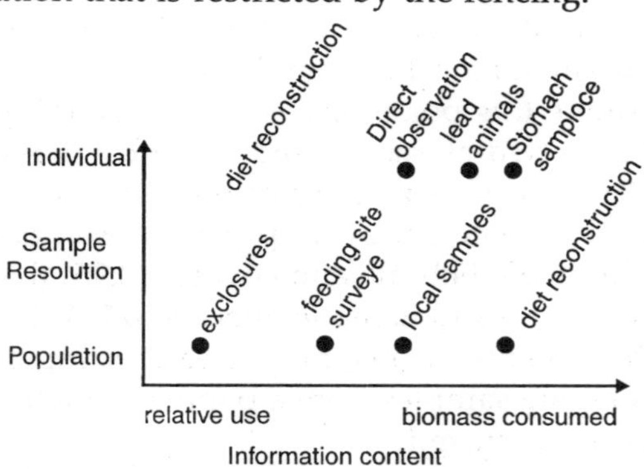

Fig. Information Content (ranging from Relative use to Estimated Biomass Consumed) and Sample Resolution (Individual Animal or Population) of Common Methods used to Investigate Vertebrate food habits.

On the other hand, samples obtained from gastrointestinal tracts can provide information on the actual biomass consumed by a specific individual of known sex and age. Studies using the latter approach are therefore much more revealing to biologists concerned with the effects of foraging patterns.

Obviously, selection of a method to investigate food use is related to the application of the subsequent results to the investigator's objectives. For example, to understand community relationships, it may be essential to partition food use patterns by age and sex of the organism being studied.

Greater sample resolution also may be needed among sexually dimorphic taxa, where a polygamous breeding system may result in spatial segregation, or among taxa that exhibit behavioral hierarchies that result in resource partition and habitat segregation.

An example using indices of diet overlap may illustrate this point. A variety of indices have been used to investigate similarity of diets or other niche parameters and often vary from 0 (no overlap or similarity) to approximately.

During an investigation of carnivore interactions, the overlap between coyotes (Canis latrans) and bobcats (Felis rufus) was 0.83. Because adult male bobcats are 50–100 percent larger than adult females, male bobcats are able to exploit larger prey and thus may be more similar to coyotes, which are substantially larger than female bobcats.

To examine this possibility, samples from bobcats were separated into two classes based on body mass and diet; overlap between each size class of bobcats and coyotes was calculated. As suspected, overlap was substantially greater between adult male bobcats and coyotes (0.95) than between coyotes and female and juvenile bobcats (0.78). This example suggests the benefits of obtaining greater sample resolution and not averaging samples from a population where resource segregation has occurred.

IMPROVING SAMPLE RESOLUTION AND INFORMATION CONTENT

There may be ways to enhance the information obtained from conventional approaches to examining food habits. Fecal samples are still the most convenient, nonintrusive method to examine food habits of vertebrates. Methods are currently available and others are being developed that may increase the information obtained from such samples. Steroid concentrations (especially estrogen) have been used to examine pregnancy rates among free-ranging mammals.

This technique could be modified to distinguish male- and female-derived fecal samples. Even greater sample resolution is possible by using emerging molecular techniques. As indicated earlier, fecal samples contain epithelial cells shed from the intestine walls of the animal depositing the sample.

DNA extracted from these cells has been used to identify the species that deposited the sample. Recently, several investigators have used this approach to identify sex and individual genetic markers. Therefore, it is possible to substantially increase the resolution of fecal samples so that researchers can track the diet of identified free-ranging individuals.

The information obtained from fecal samples could be enhanced even more by using digestibility correction factors that estimate biomass consumed. The resulting data set would probably prove very useful in evaluating diet selection and effects of consumption patterns on the forage or prey community.

As should be apparent by now, substantial information on food use patterns of vertebrates has been collected. Yet the ability of biologists to apply this information to understand factors that affect an organism's fitness or role in community structure has been limited.

Perhaps the most needed change is to ensure that future investigations have a more complete context associated with them. Rather than partitioning studies into separate efforts to examine food and habitat use, these investigations should occur (and be reported) simultaneously.

Recent advances in molecular biology will enable vertebrate ecologists to generate a more complete picture of food use patterns by specific segments of a population.

Such detailed information will enhance our ability to understand community relationships and spatialemporal patterns of vertebrate abundance. Rather than addressing general questions on the natural history of a specific species or population, clearly defined investigations of animal food habits may enhance our ability to answer the important how and why questions of vertebrate ecology.

Chapter 5

Population Density

The concept of a balance of nature goes back to the very early days of ecology. It is obvious that unlimited capacity of all animals to increase in population size or density is inevitably checked by competition for resources or the action of natural enemies. If any of these factors cause systematic changes in survival or fecundity of a population as the density increases, they are said to be density dependent.

If fecundity or survival decreases sufficiently as the population increases, then the per capita birth rate will decline to a value equal to or less than the per capita death rate and population growth will stop. In this manner, density-dependent processes constitute negative feedbacks on population growth that can maintain densities at or near an equilibrium value indefinitely.

For more than 50 years ecologists have debated whether population densities of most species are stabilized by such density-dependent factors. Howard and Fiske were the first to articulate the idea that populations cannot long persist unless they contain at least one density-dependent factor that causes the average fecundity to balance the average mortality. Other early proponents of this idea were Nicholson and Lack. In contrast, Andrewartha and Birch argued that most populations are not held at equilibrium density.

Rather, densities merely fluctuate. In their view most species avoid extinction because they comprise what we now call metapopulations. These consist of a series of subpopulations whose densities fluctuate independently of one another, but are linked by dispersal. Extinction of subpopulations occurs quite often, but these are recolonized by individuals dispersing from other subpopulations, allowing the species to persist indefinitely over the entire region. This process has been called spreading of risk. In recent years metapopulation dynamics have been explored by way of simulations that have revealed that such systems eventually go extinct in the absence of density dependence.

The debate about the ubiquity of density-dependent processes has persisted to the present day despite the efforts of various ecologists to terminate the discussion either because it was bankrupt or because they deemed that prevalence of density dependence was too obvious to deny.

Turchin provides a comprehensive review of the current status of the debate. Nevertheless, many ecologists have insisted that no conclusion regarding the existence of density dependence in a population system can be made unless their action can be demonstrated in data collected from the populations. This has proved difficult to achieve.

Until recently, adequate methods for detecting density dependence in population systems have been lacking and earlier methods have been shown to be statistically invalid. Several new methods have been proposed over the last decade, most of which involve a variety of computer-based resampling procedures.

Ecologists used to assume that populations governed by density-dependent processes had simple dynamics. They supposed that densities either remained close to equilibrium or exhibited regular oscillations about the equilibrium value. The pioneering work on deterministic chaos by Robert May taught us otherwise.

May studied the behaviour of the discrete logistic, arguably the simplest possible density-dependent population model, and showed that densities would fluctuate erratically and unpredictably if the reproductive rates were sufficiently high. Before this ecologists had assumed that the erratic fluctuations characteristic of most natural populations were caused by random influences such as weather conditions that disturbed the system away from equilibrium.

There ensued an effort to determine whether natural populations were indeed chaotic. Early studies concluded that most populations were not chaotic. These studies were based on attempts to fit natural populations to simple models and then to see whether the values of model parameters representing, for example, density dependence, time delays, or reproductive rate were such that chaotic behaviour would be expected.

The problem was that the conclusion depended on the particular model used, which was always a simplistic abstraction of the inevitably complicated dynamics of real populations. Subsequent investigators offered techniques that were more general and did not assume particular population models. Applications of these techniques have indicated that some but not all population systems are chaotic.

DETECTION OF DENSITY DEPENDENCE

ANALYSIS OF TIME SERIES OF DENSITY

For data that consist of a time series, that is, a sequence of periodic estimates of density from a population, a variety of tests have been proposed to detect the existence of density-dependent processes. If we define R as change in population density on a log scale,

$$R = \log N_{t+1} IN_t X_{t+1} - X_t$$

Where N t is density and X t is log density at time *t*, then if density-dependent processes are at work, population change should be correlated with density.

$$R = \alpha + \beta \log \mathrm{N_t} + \varepsilon_t$$

Where α and β are coefficients representing density-independent and densitydependent processes, respectively, any source of random fluctuation. The processes involved may affect fecundity, mortality, or both. If there is no density dependence, $\beta = 0$. If $\beta = 0$ it means there is no general upward or downward trend in density over time, in the absence of density dependence. Alternatively, equation can be written as

$$X_{t+1} = \alpha + (\beta + 1) X_t + \varepsilon_t$$

If both $\alpha = 0$ and $\beta = 0$, the model is known as a random walk:

$$X_{t+1} = X_t + \varepsilon_t$$

Thus to look for density dependence we could plot either R or X t+1 against X t. Under the null hypothesis of density independence we could test whether $\beta = 0$ or $\beta + 1 = 1$. Morris was the first to use plots of X t+1 against X t to search for density dependence in population systems, in his case spruce budworm populations.

Smith used this technique to show density dependence in populations of the flower thrips, which Andrewartha and Birch had used as an example of an insect governed by density independent processes. Equivalently one can plot *R* against X_t, which data on gypsy moth density collected from eight populations over a 10-year period. Various investigators have suggested solutions to the problem just illustrated. Varley and Gradwell advocated plotting X_{t+1} against X_t, interchanging them as independent and dependent variables.

Only if both regressions were significantly different from the slope of the null model ($\beta + 1 = 1$) in the same direction would density dependence be confirmed. Their approach has been subsequently shown to be extremely conservative, such that very few data sets would pass the test.

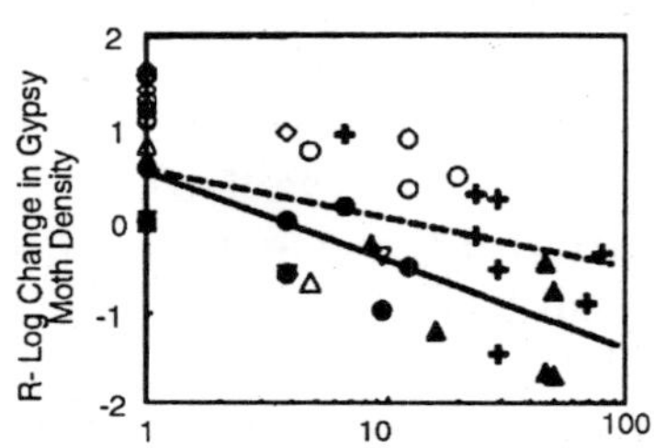

Fig. Change in Gypsy Moth Density (R = log 10 [N t+1 /N t]). Plotted Against Gypsy Moth Density (N t) Compared to Slope (dotted line) expected with no density dependence. Solid line is Regression Equation fit to the Data.

Values of $R < 0$ indicate declining densities; values of $R > 0$ indicate increasing densities. Densities are estimated egg masses per hectare.

Pollard et al. proposed a so-called randomization test that has emerged in several comparisons as the most accurate and powerful of extant methods. They calculated a null distribution of slopes (i.e., those expected if there were no density dependence) to compare against the slope obtained from the actual data. This was done by generating a large number (e.g., 1,000) of permutations of the list of successive annual census data values.

Because the data values were scrambled in this way, the dependence of each value on the preceding value was obliterated, yet the overall variance of the densities was retained. For each permutation they regressed R versus X t and then compared the slope from the actual data with the distribution of slopes obtained from the 1,000 permutations. Reddingius proposed a very similar test.

Dennis and Taper offered a procedure known as the parametric bootstrap likelihood ratio test that was similar to that of Pollard et al. Instead of permuting the data they generated 1,000 simulated data sets under the null model of density independence with randomly chosen error terms.

As Pollard et al. had done, they then calculated a distribution of slopes with which they compared the actual slope generated from the data by ordinary linear regression. After applying both of these techniques to the data given. Three of eight population time series were identified as density dependent (at $p < 0.05$) with the test of Dennis and Taper and two out of eight with the test of Pollard.

However, neither of these techniques is without problems; first they all lack statistical power. One needs 20–30 generations of data to reliably find density dependence when it exits. Data sets that long are rare in ecology. Second, Shenk et al. simulated the effect of measurement error on both of these tests.

They concluded that both tests were highly prone to type I error (concluding density dependence when it did not exist) and hence of little use when measurement error is significant, as it usually is in most data sets. Dennis and Taper conducted analogous simulations and found that their test was robust against measurement error.

The difference in the two studies was in how measurement error was modeled and it is too early to tell which view will prevail. Finally, correlations between error terms from one year to the next can also lead to spurious conclusions of density dependence when it does not exist.

It would not be very surprising to have autocorrelated errors because they include the effects of all other variables on population growth other than density. Suppose a population was determined largely by the action of a generalist predator whose density was not linked to that of its prey and that might or might not cause density-dependent mortality.

Variation in predation rates caused by fluctuation in predator density would be embodied in the error term and would probably be influenced by predator densities at previous time steps. The error term would thus be autocorrelated.

ANALYSIS OF DATA ON MORTALITY OR SURVIVAL

Many investigators collect data not just on density but on mortality or fraction surviving in particular age categories or life stages. They may be interested in particular agents of mortality and want to know whether these are capable of stabilizing the population. Plots of some measure of percentage mortality versus density of the population reveal density dependence.

For example, Varley and Gradwell presented data on the winter moth, a defoliator of oak trees in Great Britain. They collected data on sources of mortality on successive instars or life stages over a period of 15 generations (years) at one site. They expressed mortality as k-values (k = —log 10 (proportion surviving)) and plotted k-values for each cause of mortality against the log density of the individuals present at the beginning of the stage or age category on which the agent of mortality acted. They used standard linear regression to determine whether mortality increased or decreased with density. The strong negative bias described earlier for analysis of density time series was not present here because the measurements of mortality differed from the measures of density.

However, a number of statistical problems involve violations of the usual assumptions of linear regression. The regression may be nonlinear, measurement error may affect the estimates of both density and mortality, the variance of the k-values may vary systematically with density, and the error terms may not be independent because the data are obtained from time series.

Solutions have been proposed for several of these problems. However, Vickery analyzed the various extant methods and found them all either biased or lacking in statistical power. He advocated using a randomization test identical to that of Pollard et al. but applied to data on stage-specific mortality instead of time series of density.

Use of these techniques rests on the assumption that the data collected for mortality and density are accurate and unbiased. This may not be easy to achieve, particularly when ages or life stages overlap temporally. Various techniques may be used to convert densities or numbers present in periodic samples to estimates of numbers entering particular life stages or age categories.

Similar techniques exist to convert mortalities or rates of infection obtained from periodic samples to the stage-specific mortality that best represents the overall impact of the agent of mortality. Additional techniques may be required when two or more agents of mortality act contemporaneously. It is beyond the scope of this review to describe these techniques here.

Density dependence is often reported in studies in which data on mortality or survival are obtained simultaneously from several different populations that vary in density. It is important to realise that the processes that give rise to density dependence in such studies may not be the same as those producing density dependence in studies wherein mortality and density are shown to vary over time from one or more populations. For example, Gould et al. showed that mortality of gypsy moths caused by a parasitic fly, Compsilura concinnata, increased dramatically with gypsy moth density on a series of experimental populations created with different densities at several locations in the same year.

The density-dependent response was evidently a behavioral one by the fly. It was not at all clear the extent to which such responses would occur in studies in which density varied temporally instead of spatially. Only in the latter studies would the reproductive response of the fly to changes in gypsy moth density be measured. Indeed, a 10-year study of parasitism in naturally occurring populations of gypsy moth revealed no evidence of temporal density dependence and far lower levels of parasitism by C. concinnata. The ability of this fly to regulate low densities of gypsy moth is thus questionable.

Several investigators have surveyed the published literature on life tables in particular taxa in order to ascertain how often density dependence has been detected.

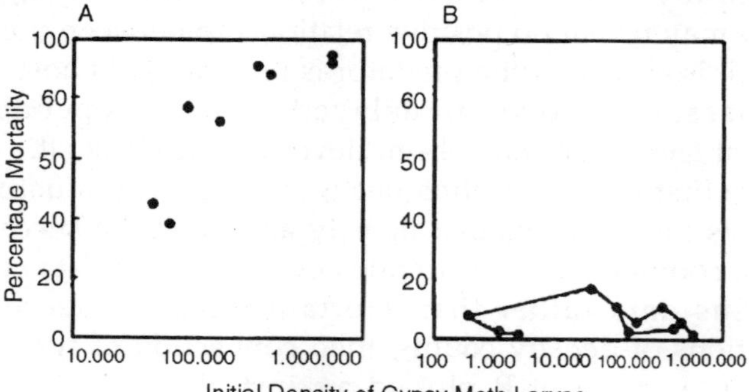

Fig. (A) Percentage mortality of gypsy moth caused by the parasitic fly Compsilura concinnata to a series of experimental populations created with different densities in the same year. (B) Time series of percentage mortality caused by C. concinnatain a 10-year study of gypsy moth in naturally occurring populations. Solid line in (B) connects consecutive generations.

A typical conclusion is that evidence of density

dependence is rare. Part of the reason for this is that most life table studies are of short duration and the problems of statistical power indicated earlier limit our ability to detect density dependence when it exists.

In addition, data from most studies contain a considerable degree of measurement error. Such error can also obscure density-dependent relationships. Finally, the action of many density-dependent mortalities may lag behind those of their hosts. The methods described earlier will not detect their action.

DETECTION OF DELAYED DENSITY DEPENDENCE

Time lags in density-dependent responses are common in population systems. For example, it is typical for a predator or parasitoid to respond numerically to changes in density of its host, but this response typically lags behind that of its host by at least one generation.

The result is that peak predator density and hence peak mortality of the host occurs after the host has declined dramatically from peak density. Plots of mortality against density may reveal no positive relationship between the two, even if it is clear that the predator is regulating its host. Such responses are known as delayed density dependence. Different techniques have been developed to detect it.

The first of these techniques were graphical in nature. If one plots mortality against density and connects successive years, a counterclockwise spiral is evident. If the data consist of census data rather than mortality, that is, successive generations of density counts, then connection of successive years on a graph of R plotted against N t or X t yields a clockwise spiral.

A major advance in detection of delayed density dependence was developed by Turchin. He applied time series analyses that have had wide application in econometrics and the physical sciences. The methods involve fitting a model similar to equation 6.2 but with terms representing the effects of density in generations before the last one: Partial autocorrelation analysis tells you whether there is significant delayed density dependence.

The first one is undoubtedly the best known time series in ecology, the snowshoe hare and lynx oscillation in Canada based on pelts delivered to the Hudson Bay Company over a time period exceeding 100 years.

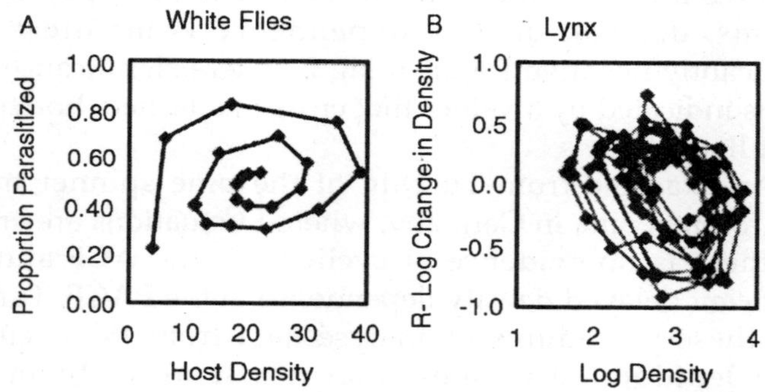

Fig. Graphic Detection of Delayed Density Dependence Wherein Points Representing Consecutive Generations are Connected.

(A) Percentage mortality caused by the parasitoid Encarsia formosa on the greenhouse white fly (Trialeuroides vaporarorium).

These data have been analyzed by Royama. Fluctuations in density of the lynx populations, along with the autocorrelation function (ACF) and the partial autocorrelation function (PACF). The ACF expresses the correlation between each measure of density and the densities 1, 2, ... n generations back (the lag time). For time series that display pronounced cycles, the ACF reaches a peak value at a lag time corresponding to cycle period.

The ACF can be used to determine whether time series are truly cyclic. Although the cycles are obvious in many time series the fluctuations are more irregular and detecting the difference between cyclic behaviour and random fluctuations is not at all obvious. Delayed density dependence can be detected by looking at the PACF.

The PACF represents the correlation that remains at each lag time (between X t and X t"n), with the correlations due to smaller lag times removed. For example, in most population time series there is a fairly high correlation between X t and X t"1. For the same reason X t"1 is correlated with X t"2, and, consequently, a positive correlation exists between X t and X t"2, albeit a weaker one.

The PACF removes these effects due to shorter lags and measures the correlation that remains. For most population systems, delayed density dependence is manifest in a significantly negative PACF at lag 2, where significance ($p < 0.05$) is indicated by a values that cross the dashed, horizontal cutoff lines.

An example from Turchin of the pine spinner moth, Dendrolimus pini, in Germany, where fluctuations are erratic and there is no evidence of cyclicity in the ACF and yet significant delayed density dependence in the PACF. Turchin used these procedures to analyse data from 14 species of forest lepidoptera to show that nearly all of them had significant lag 2 or higher effects.

In contrast, Hanski and Woiwod analyzed 5,715 annual time series of moths and aphids captured in survey traps. They found a high incidence (67–91 percent) of density dependence but less than the 5 percent incidence of delayed density dependence they would have expected by chance alone.

Holyoak suggests that part of the unexpectedly low incidence of delayed density dependence may be caused by the multiple generations that elapsed between the annual samples for many of these species in contrast to the forest lepidoptera analyzed by Turchin.

Several of the same limitations of tests for direct density dependence also apply to tests for delayed density dependence, The techniques have little or no statistical power for the short time series that are typical of most ecological data. Furthermore, autocorrelations in the error terms can lead to spurious conclusions of delayed density dependence.

Berryman and Turchin argue that Williams and Liebhold's conclusion is overly pessimistic and based on simulations using unrealistic parameter values. Williams and

Liebhold reply that the parameter values were typical of populations analyzed earlier by Turchin.

DETECTION OF CAUSES OF POPULATION CHANGE

KEY FACTOR ANALYSIS

Many population ecologists are more interested in determining the causes of changes in density than in the causes of stability. For example, ecologists who work with animals that exhibit outbreak dynamics may want to uncover the cause of these outbreaks.

Key factor analysis was developed to identify such causes, or at least the life stage on which they act. Like the density dependence techniques discussed earlier, key factor analysis applies to population systems for which data exist on survival or mortality of specific ages or life stages in a population over a series of consecutive generations.

The first methods developed were graphic ones. Varley and Gradwell advocated plotting mortalities in each life stage expressed as k-values along with total generational mortality (K = X t — X t+2) against generation number (time). They included changes in fecundity as a k-value so total K represents total generational change in density (K = —R) and is the sum of n sequential, stage-specific k-values (K = k 1 + k s +.. k n). The key factor was the one whose fluctuations most closely matched those of total K.

For example key factor analysis of a 10-year study of partridge populations in England (Blank et al. 1967). Mortality (or loss of fecundity) in this population was attributed to eight sequential causes. Of these, mortality of chicks (k 4) was the one whose fluctuations clearly matched of that of total generational change (K).

For some species such graphic analyses might not yield a definitive answer. Podoler and Rogers advocated calculating regression lines of each kvalue against total K. The key factor was the one with the largest positive slope. They applied this technique to a number of data sets, including the English partridge data. Chick mortality (k 4) had the steepest slope, thus confirming the conclusion of the earlier graphic analyses. Podoler and Rogers recognized that one could not test for the significance of the slope in the usual way because the axes in the regression were not independent.

Manly offered a more definitive analytic approach based on partitioning the variance of R into its additive components and constructing a variance—covariance matrix of all the k-values or causes of mortality. The key factor was the mortality or life stage with the largest variance component and was not always the same as that obtained with the earlier graphic methods.

Manly applied his technique to the partridge data given earlier and concluded that k 7 (losses in late winter due to natural causes) was the key factor. Whereas k 4 accounted for much of the variation in early season mortality, most of it was compensatory to the earlier mortality and thus not the main cause of population change.

Key factor analysis was originally designed for univoltine insects that reproduced during a short-lived adult stage and survival data were confined to preadult stages. Problems arise when this technique is applied to organisms, including most vertebrates, for which reproduction extends over a substantial fraction of the typical life.

Survival during the oldest age classes weighs equally with younger age classes in key factor analyses, yet older age classes may contribute little or even nothing to the reproduction of the next generation. Brown et al. offered a technique they called structured demographic accounting that was designed to solve this problem and others arising from key factor analysis. This method was similar to that of Manly in that the total variation in R was decomposed into the variances and covariance of the component recruitment (births) and survivals during particular age classes.

However, this process was done separately for each of several age classes so that variation in births as well as deaths due to each class could be properly assessed. A modification of this approach has been proposed by Silby and Smith, who advocated calculating the impact of each k-value on population growth rate (r;er = R) rather than generational survival. Population growth rate is measured by standard life table methods, which account for age-specific fecundity and survival.

The limitations of key factor analysis have been documented by various authors. Kuno showed that sample error and compensatory densitydependent mortality can lead to spurious conclusions in key factor analyses. Manly acknowledges these as important limitations.

Royama identified several further problems. One of these was that the factors causing population change may fluctuate on very different time scales and those responsible most of the variation in R in a set of data may not be the factors responsible for the onset or decline of high-density conditions for species prone to outbreaks. Royama illustrated this with an example from spruce budworm, a major defoliator of forest trees in Canada.

The rise and eventual decline of a population over a 10-year period were caused by a steady decline in larval survival, whereas recruitment (egg laying) fluctuated more from year to year and was identified by the techniques described earlier as the key factor. This example illustrated that key factor analyses may give misleading results or may not be easy to interpret. A single analytic process such as key factor analysis may not suffice to unravel the causes of density change.

EXPERIMENTAL MANIPULATION

Many ecologists have turned to experimental manipulations to provide proofs of the effects of various factors on population density or survival. For example, it is possible to establish experimental populations that differ in density and then to measure the mortality cause by various natural enemies.

In (A), Generational Change (K= —R) and Individual Sources of Mortality are Expressed as k-values: (k =—log (survival)) for each of eight sequential causes is plotted for each year of the study. In (B) and (C) k-values are regressed against generational change (total K). As indicated earlier, mortalities may vary in density spatially between plots, but these need not be the same as those that vary temporally in the same plots. Application of the techniques described early is appropriate only where the data represent time series.

However, there is no reason why the time series could not be obtained from populations that are experimentally manipulated. A well-known example of such a study is the kilometre-scale exclosure study of the factors causing fluctuations in snowshoe hares in Yukon Territories.

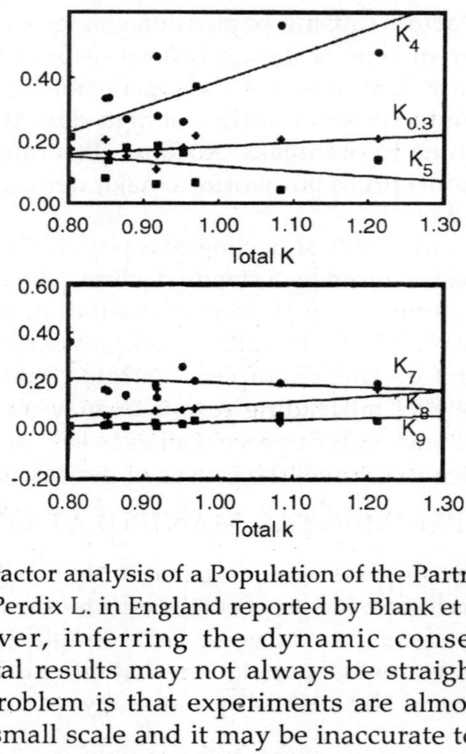

Fig. Key Factor analysis of a Population of the Partridge Perdix Perdix L. in England reported by Blank et al.

However, inferring the dynamic consequences of experimental results may not always be straightforward. A common problem is that experiments are almost inevitably done on a small scale and it may be inaccurate to extrapolate to the larger scale of natural populations. For example, Gould et al. manipulated gypsy moth density on 1-ha plots and showed that tachinid parasitoids decimated populations in a strongly density-dependent way.

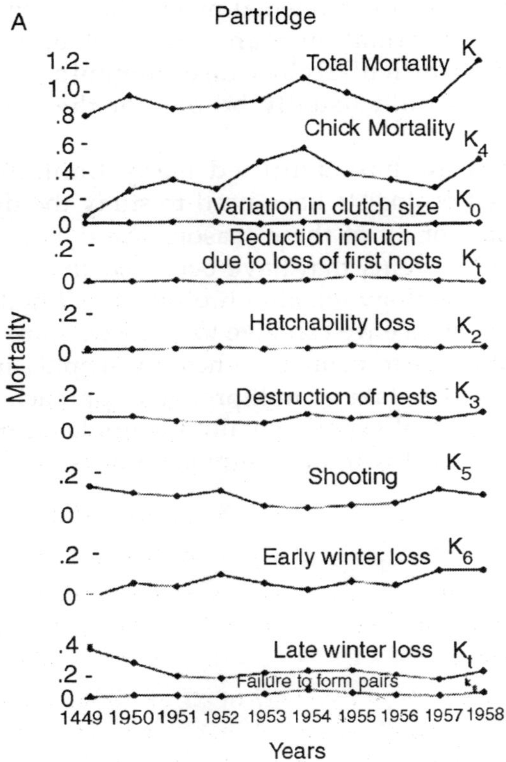

However, studies from natural populations revealed far lower levels of parasitism caused by these species and scant evidence for density dependence. Why the difference? Evidently the parasitoids aggregated to the 1-ha experimental populations from surrounding areas of low gypsy moth density. This density-dependent aggregation response would be nullified in natural populations, where densities rise simultaneously over much larger areas.

The spatial scale in this study was thus crucial. Manipulation on a 1-ha spatial scale, although large relative to most ecological experiments, was not large enough to mimic the dynamics of natural populations. In other words, experimental manipulations may introduce a variety of artifacts that may be difficult to detect.

Nevertheless, experimental manipulation almost always yields more information than studies of unmanipulated populations, particularly because unmanipulated control populations would usually be part of the experimental design.

This review has identified many limitations in the methods that ecologists have used to study the dynamics of populations. For all of these reasons the dynamics of even well-studied systems that have occupied the talents of the best minds in ecology remain unresolved and hotly debated.

The best advice we can give to those who are embarking on such studies is to maintain a healthy skepticism of all the techniques and to take a multipronged approach. Wherever possible, studies of experimentally manipulated populations should be coupled with unmanipulated ones.

Chapter 6

Monitoring Populations

Assessing changes in local populations is the key to understanding the temporal dynamics of animal populations, evaluating management effectiveness for harvested or endangered species, documenting compliance with regulatory requirements, and detecting incipient change. For these reasons, population monitoring plays a critical role in animal ecology and wildlife conservation.

Changes in abundance are the typical focus, although changes in reproductive or survival rates that are the characteristics of individuals, or other population parameters, also are monitored. Consequently, many researchers and managers devote considerable effort and resources to population monitoring. In doing so, they generally assume that systematic surveys in different years will detect the same proportion of a population in every year and changes in the survey numbers will reflect changes in population size.

Unfortunately, these assumptions are often violated. In particular, the following two questions are pertinent to any animal ecologist involved in population monitoring. First, is the index of population abundance used valid? That is, does variation in, for example, track densities of mammals, amphibian captures in sweep nets, or counts of singing birds reliably reflect changes in local populations of these organisms?

Second, does the design of a monitoring programme permit a reasonable statistical probability of detecting trends that might occur in the population index? In other words, are estimates of population indices obtained across a representative sampling of habitats and with sufficient intensity over time to capture the trends that might occur in the population being monitored? Failure to address these questions often results in costly monitoring programs that lack sufficient power to detect population trends.

ABUNDANCE RELATIONSHIPS

TYPES OF INDICES

Making accurate estimates of absolute population size is difficult. Animals often are difficult to capture or observe, they are harmed in the process, or the associated costs and effort of making absolute counts or censuses are prohibitive. Therefore, animal ecologists often rely on indices of population size and monitor these indices over time as a proxy for monitoring changes in actual population size.

Indices may be derived from sampling a small fraction of a population using a standardized methodology, with index values expressed as individuals counted per sampling unit (e.g., fish electroshocked per kilometre of shoreline, tadpoles caught per net sweep, salamanders captured per pitfall trap, birds intercepted per mist net, or carcasses per kilometre of road). These examples involve direct counts of individuals.

When individuals of a species under study are difficult to capture or observe, another class of indices makes use of indirect evidence to infer animal presence. Auditory cues are often used as indirect indices (e.g., singing birds per standard listening interval, overall sound volume produced by insect aggregations, howling frequency by packs of wild canids, or calling intensity in frog choruses).

Other indirect indices are based only on evidence of animal activity (e.g., droppings per unit area, tracks per unit transect length or per bait station, or quantity of food stored per den).

ABUNDANCE FUNCTIONS

An index to population size (or abundance) is simply any "measurable correlative of density" and is therefore presumably related in some manner to actual abundance. Most animal ecologists assume that the index and actual abundance are related via a positive, linear relationship with slope constant across habitats and over time.

In some situations, these relationships hold true. However, the relationship often takes other forms in which changes in the index may not adequately reflect changes in the actual population.

A nonlinear (asymptotic) relationship may be common in situations where the index effectively becomes saturated at high population densities. Such may be the case for anurans monitored using an index of calling intensity. The index is sensitive to changes at low densities of calling male frogs in breeding choruses because calls of individuals can be discriminated by frog counters.

At higher densities, however, calls of individual frogs overlap to an extent that size variation of choruses cannot be discriminated by observers. In other words, the index increases linearly and positively with abundance to a threshold population density, and then becomes asymptotic.

Another example of a nonlinear index—abundance relationship concerns use of presence/absence as a response such that the proportion of plots occupied by a given species is the index of abundance. At low population densities, changes in population size can be reflected in changes in degree of plot occupancy. Once all plots are occupied, however, further population increases are not reflected by the index because the index becomes saturated at 100 percent occupancy.

A final example involves bait stations for mammals, which may be frequented by subdominant animals more at low population densities than at high densities because of behavioral inhibition. The main implication of this type of nonlinear index—abundance relationship is that it prevents detection of population change (in any direction) above the saturation point of the index.

A threshold relationship also may occur in index—abundance relationships if the index effectively bottoms out at low population densities. For example, if sample plots are too small, listening intervals too short, or sample numbers too few, observers may simply fail to register individuals even though they are present at low densities.

Consequently, detection of population change below the threshold of the index is precluded. This situation probably occurs in surveys for many rare, endangered, or uncommon species. The threshold and saturation phenomena can combine in some situations.

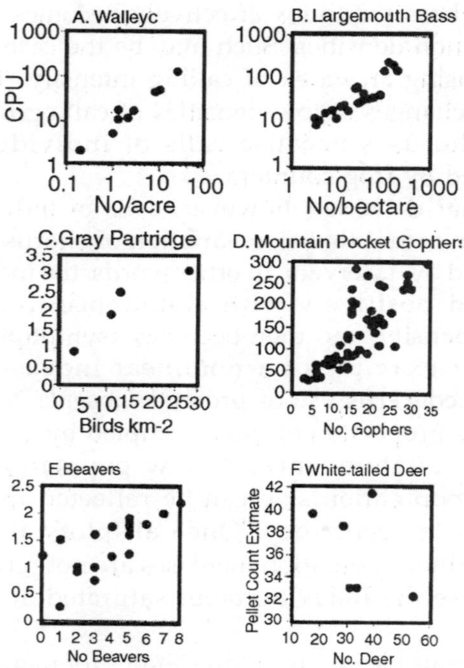

Fig. Relationship between Population Indices (vertical axis) and Actual animal Abundance (horizontal axis).

(A) From Serns (1982),
(B) From Hall (1986),
(C) From Rotella and Ratti (1986),
(D) From Reid et al. (1966),
(E) From Easter-Pilcher (1990),
(F) From Ryel (1959).

For example, because calling behaviour may be stimulated by group size in frogs, individuals may not call (or may do so infrequently) when choruses are small and may be overlooked by frog counters, but increasing numbers of calling frogs above a certain threshold may also be indistinguishable to frog counters.

Occasionally indices used have no relationship to abundance, although sometimes an apparent lack of an index—abundance relationship may simply be a result of sampling error or too few samples taken to verify the relationship.

Nevertheless, the possibility that a seemingly reasonable, readily measured index has no relationship to the actual population must always be considered by animal ecologists using an unverified index, and preferably be examined as a null hypothesis during a pilot study.

VARIABILITY OF INDEX

Independent of the specific form of the index—abundance relationship, most researchers assume it to be constant among habitats and over time. However, in perhaps the most comprehensive validation study of an indirect index, a study by Reid et al. on mountain pocket gophers (Thomomys talpoides), the index used (numbers of mounds and earth plugs) consistently displayed a positive, linear relationship to actual gopher numbers, whereas the intercept and slope varied substantially between habitats.

Other situations, such as electroshocking freshwater fishes, apparently yield comparable index—abundance relationships between habitats despite large differences in densities between habitats. In contrast, index—abundance relationships in different habitats can be reversed although these examples may be compromised by sampling error. Finally, the slope, intercept, and precision of the relationship may vary among years within the same habitats.

Inferences about population change drawn from indices are also often hampered by sampling error. Whatever the form of the index—abundance relationship between habitats and over time, the precision of the relationship can be quite low. This is particularly true for indirect indices, in which variation is strongly influenced.

By environmental factors such as weather and time of day, as well as by observers. Such index variation can substantially reduce the power of statistical tests examining changes in index values between sites or over time.

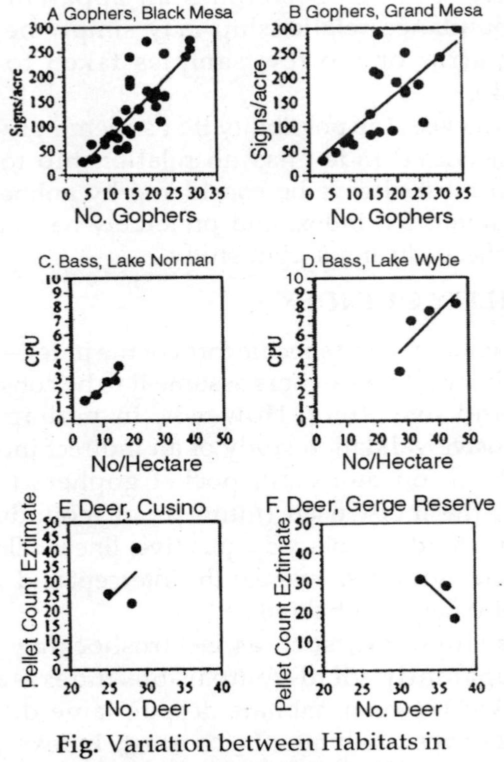

Fig. Variation between Habitats in Index—abundance Relationships.

(A) and
(B) From Reid et al.,
(C) and
(D) from McInerny and Degan,
(E) and
(F) from Eberhardt and Van Etten.

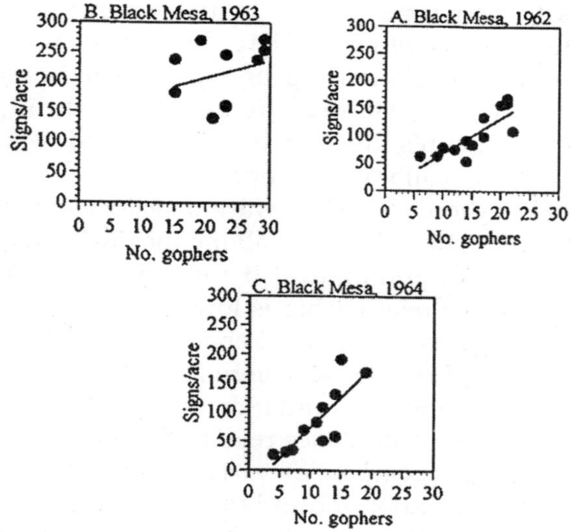

Fig. Variation in the Index— abundance Relationship over time at the same site.

IMPROVING INDEX SURVEYS

The few studies attempting to validate indices suggest that population indices and absolute abundances are rarely related via a simple positive, linear relationship with slope constant across habitats and over time. Thus animal ecologists would do well to proceed cautiously when designing and implementing index surveys.

In particular, index validation should be considered a necessary precursor to implementing index surveys. Some guidance on the relationship of the index to abundance may be found in the literature, but index validation studies are rare. Lacking such information, conducting a pilot study using the index in areas where abundance is known or can be estimated is useful.

Such a validation study would need to be replicated across multiple sites that exhibit variation in population size or density, or over time at a site where abundance varies over time. Making multiple estimates of the index:abundance ratio at each site and time period is also useful so that the contribution of sampling error to the overall noise in the index—abundance relationship among sites can be estimated. Validation studies also may be advisable throughout a monitoring program's life span because the index may need to be periodically calibrated or updated.

Ecologists should also be aware that developing indices that have a 1:1 relationship with abundance will most reliably reflect changes in abundance. If the slope describing the index—abundance relationship is low, then large changes in abundance are reflected in small changes in the index. Such small changes in the index are more likely to be obscured by variation in the index—abundance relationship than if the slope of the index—abundance relationship were higher.

Methods of reducing index variability and increasing the precision of the index—abundance relationship include adjusting the index by accounting for auxiliary variables such as weather and observers. In practice, these factors may be overlooked if many years of data are gathered because the short-term bias they introduce typically is converted simply to error in long-term data sets.

In an ideal situation, each index would be validated, adjusted for sampling error by accounting for external variables, and corrected to linearize the index and make it comparable across habitats and over years. However, this is rarely an option for regional-scale surveys conducted across multiple habitats over many years by many people and involving multiple species, although it may be possible for local monitoring programs focused on single species.

The following advice may be useful to animal ecologists for improving index surveys. First, the basic relationship between the index and abundance should be ascertained to determine whether the index might yield misleading results and therefore should not be implemented. Second, any results from trend analysis of index data should be considered in light of potential limitations imposed by the index—abundance relationship.

For example, saturated indices could be the cause of a

failure to detect population changes. Most importantly, animal ecologists must be cautious about concluding that a lack of trend in a time series of index data indicates population stability. Often an index may be unable to capture population change because of a flawed index—abundance relationship or simply excessive noise caused by sampling error in the index.

SPATIAL ASPECTS OF MEASURING CHANGES IN INDICES

Many animal ecologists are concerned with monitoring multiple local populations with the intent of extrapolating changes observed in those populations to larger, regional populations. In such a case, the sample of areas monitored must be representative of areas in a region that are not sampled if observed trends are to be extrapolated to regional populations. Selection of sites for monitoring is therefore a key consideration for animal ecologists concerned with identifying change in regional populations.

Balancing sampling needs and logistical constraints in the design of regional monitoring programs can be problematic, however. For sampling areas to be representative, random selection of sites for surveying is advised, but a purely random scheme for site selection is often unworkable in practice.

For example, sites near roadsides and those on public lands are generally easier to access by survey personnel than are randomly selected sites. Also, monitoring sites that occur in clusters minimize unproductive time traveling among survey sites. Time is generally at a premium in monitoring efforts not only because of the costs of supporting survey personnel but also because the survey window each day or season for many animals is brief.

A simple random sample of sites may also produce unacceptably low encounter rates for the organisms being monitored (too many zero counts to be useful). This could be overcome by stratifying sampling according to habitat types frequented by the species being monitored.

However, information on habitat distributions in a region from which a stratified random sampling scheme might be developed often is not available to researchers. Furthermore, prior knowledge of habitat associations of most species that can be used as a basis for stratification often is not available. Finally, ecologists often monitor multiple species for which a single optimal sampling strategy may simply not be identifiable.

These difficulties in implementing random sampling schemes imply that nonrandom site selection schemes may be the most practical way to organize sampling for monitoring programs. However, animal ecologists would do well to be aware of the serious and lasting potential consequences of nonrandom site selection.

Researchers initiating a survey programme are often drawn to sites with abundant populations, where counts are initiated under the rationale that visiting low-density or unoccupied sites will be unproductive. If the populations or habitats under study cycle, however, then initial counts may be made at cycle peaks.

As time progresses, populations at the sites selected will then tend, on average, to decline. The resulting pattern of decline observed in counts is an artifact of site selection procedures and does not reflect any real population trend.

This sampling artifact can lead researchers to make erroneous conclusions about regional population trends. This problem has compromised a regional monitoring programme for amphibians and regional game bird surveys. These examples highlight why site selection can be an important pitfall in designing monitoring programs.

Unfortunately, few simple recommendations can be made for guiding the process. A detailed knowledge of habitat associations of the species under study, as well as the distribution of those habitats in a region, can provide useful guidance to animal ecologists in selecting a sampling design that is logistically feasible to monitor.

Stratifying (or blocking) sampling effort based on major habitat features such as land cover type will almost always yield gains in precision of population estimates each sampling interval.

Specifically, researchers would do well to identify species—habitat associations and generate regional habitat maps before initiating surveys so that the explicit tradeoffs between alternative sampling schemes, logistical costs, and sampling bias can be evaluated.

One workable solution to this problem involves two steps. First, populations at selected sites that are presumably representative of particular habitat strata in a region are rigorously monitored. Second, an independent programme is established that explicitly monitors changes in the distribution and abundance of habitats in the region. Trends in habitats can then be linked to trends in populations at specific sites to extrapolate regional population trends.

MONITORING INDICES OVER TIME

Once animal ecologists attempting to monitor populations have addressed issues of index validity and sampling schemes for selecting survey sites, another set of issues related to the intensity of monitoring over time must be considered.

These issues include how many plots to monitor, how often to survey plots in any given year, the interval and duration of surveys over time, the magnitude of sampling variation that occurs in abundance indices, and the magnitude of trend variation in local populations in relation to overall trends in regional populations.

Other less obvious but often equally important factors to be considered include β levels and desired effect sizes (trend strengths) set by researchers. Specifically, researchers need to specify the probabilities at which they are willing to make statistical errors in trend detection, that is, the probability of wrongly rejecting the null hypothesis of no trend (at a probability = β, that is, the level of significance) and of wrongly accepting the null hypothesis of no trend (at a probability = α).

Furthermore, the statistical method chosen to examine trends in a count series also can influence the likelihood of detecting them. Understanding how these factors interact with the inherent sampling variation of abundance indices can provide insights into the design of statistically powerful yet labour-efficient monitoring programs.

Statistical power underlies these issues and provides a useful conceptual framework for biologists designing studies that seek to identify population change. The key problem identifying population change is that sources of noise in sample counts obscure the signal associated with ongoing population trends.

Trends represent the sustained patterns in count data (the signal) that occur independently of cycles, seasonal variations, irregular fluctuations that are sources of sampling error (the noise) in counts. Statistical power simply represents the probability that a biologist using a particular population index in conjunction with a specific monitoring protocol will detect an actual trend in sample counts, despite the noise in the count data. In a statistical context, power is the probability that the null hypothesis of no trend will be rejected when it is, in fact, false, and is calculated as $1 - \alpha$.

Although statistical power is central to every monitoring effort, it is rarely assessed. Consequences of ignoring power include collecting insufficient data to reliably detect actual population trends. Occasionally, collection of more data than is needed occurs. Unfortunately, until recently few tools have been available to animal ecologists that permit assessment of statistical power for trends.

POWER ESTIMATION FOR MONITORING PROGRAMS

The large numbers of factors that interact to determine the statistical power of a monitoring programme make power estimation a complex undertaking. Analytical approaches are forced by the large number of variables involved to over simplify the problem. Because of the complexities involved in generating power estimates for monitoring programs, the problem may be most tractable with simulation methods.

Accordingly, a conceptually straightforward Monte Carlo approach based on linear regression analysis has been devised. With this approach a researcher defines the basic structure of a monitoring programme and provides a variance estimate for the population index used. Simulations are then run in which many sets of sample counts are generated based on the structure of the monitoring programme with trends of varying strength underlying them.

The frequency with which trends are detected in the counts, despite the sampling error imposed by the population index and the structure of the monitoring programme, reflects the power of the monitoring design to detect trends. The simulation programme is particularly useful for evaluating the tradeoffs between sampling effort, logistical constraints, and power to detect trends.

VARIABILITY OF INDICES OF ANIMAL ABUNDANCE

A key influence on power to detect a given population trend is the variability of the population index used. Power to detect trends is inversely related to the magnitude of index variability and monitoring programs must be designed around the component of index variability that cannot be controlled.

In other words, sufficient numbers of plots must be monitored frequently enough to capture trends despite the inherent variability of the population index. Without pilot studies, however, researchers often have no estimate of population variability. Lacking estimates of this critical parameter impairs the ability of animal ecologists to design statistically powerful monitoring programs.

A ready source of data on the variability of population indices can be found in published time series of population counts. Hundreds of long-term population studies for a variety of taxa have been published in the last century, albeit mostly for temperate-zone organisms.

Because most of these population series were generated using population indices, not population censuses, presumably variation in these count series reflects both environmental variation in the populations and sampling error associated with the counting methodology. As long as the time series are of sufficient and comparable duration, significant trends have been removed from them, and sufficient numbers of studies have been made, approximations of index variability can be estimated.

Further- more, these estimates can be integrated with power analyses to provide general guidance on sampling protocols that animal ecologists can use to design robust monitoring programs for local populations.

To this end, count series of local animal and plant populations that extended more than 5 years were obtained by examining 25 major ecology journals published from 1940 to the present (nonwoody plants are also presented here because animal ecologists often must monitor plant populations in the course of their animal studies).

Variability of each count series thus obtained was estimated by dividing the standard deviation of the counts by the mean count to determine the coefficient of variation (CV). To remove trends in the counts (which might have inflated variance estimates), the standard deviation was determined from the standardized residuals of a linear regression of counts against time.

Furthermore, because the variability of a time series is related in part to its length, a 5-year moving CV (similar in concept to a moving average) was calculated for each count series. (However, most studies of birds, moths, and butterflies failed to present raw counts that could be detrended and standardized, so the means and error terms as presented in these studies were used.

The index variabilities for these groups are therefore potentially biased high in relation to those estimates for other taxa). CVs were subsequently averaged within groups of taxonomically and ecologically related species.

A total of 512 time series for local animal and plant populations were analyzed, which provided estimates to calculate average index variabilities for each of 24 separate taxonomic and ecological groups. Few groups had low variability indices (CV below 25 percent), including large mammals, grasses and sedges, and herbs. A larger number had intermediate variability indices (CV 25–50 percent), including turtles, terrestrial salamanders, large birds, lizards, salmonid fishes, and caddis flies.

Most groups had indices with CVs between 50–100 percent, including snakes, dragonflies, small-bodied birds, beetles, small mammals, spiders, medium-sized mammals, nonsalmonid fishes, pond-breeding salamanders, moths, frogs and toads, and bats.

Finally, only butterflies and drosophilid flies had average indices with CVs above 100 percent. Although a pilot study is clearly preferable, lacking one of their own animal ecologists can refer to the specific studies or to the summary for information useful for designing monitoring programs for a particular species.

It is important to note that index variabilities reflect temporal variation inherent in populations as well as sampling error associated with the counting methods. For example, direct count methods were used most often for those groups with the lowest index variability, including large mammals, all plants, terrestrial salamanders, and large-bodied birds. An exception was butterflies, which typically were counted with time-constrained visual searches.

Nets and traps were used to capture individuals in most remaining groups. Trapping methods that sampled only a segment of a population (e.g., frogs, toads, and pond-breeding salamanders on breeding migrations) or that relied on attractants (e.g., most small- and medium-sized mammals at bait stations, moths and caddis flies at light traps, and drosophilid flies at fruit baits) were associated with high index variabilities.

Similarly, most studies of small-bodied birds were based on counts of singing individuals and also displayed high variability. Both method-associated sampling error and inherent population variability clearly make important contributions to overall index variability, and the recommendations that follow assume that researchers will use the same standardized counting methods used by the researchers who generated the count series analyzed here.

SAMPLING REQUIREMENTS FOR ROBUST MONITORING PROGRAMS

Estimates of index variabilities were incorporated into a power analysis to generate sampling recommendations for animal ecologists for designing effective programs for monitoring local populations. The power analysis assumed the following logistical constraints.

Resources available for a local or regional monitoring programme would permit surveys of up to 500 plots or subpopulations on one to five occasions annually over a monitoring period of 10 years. Average plot counts for all groups were assumed to equal 10, with count variances comparable to the average value calculated for each group based on the literature survey.

Trends in the population index were assumed to be linear, β and α were set at 0.05, and tests of significance were two-sided. Within this framework, sampling requirements to detect overall changes in population indices of 10 percent, 25 percent, and 50 percent for each group were estimated.

This analysis indicated that infrequent monitoring (for example, once or twice per year) on a small number of sites or plots (10 or less) would reliably detect strong population trends (that is, a 50 percent change over 10 years) in most groups. Even for highly variable groups frequent monitoring (three to five times per year) of a small number of plots (30 or less) would permit detection of a trend of this magnitude.

However, more intensive monitoring is needed to detect weaker trends of 25 percent and 10 percent, but nevertheless is still at a logistically feasible level (100 or fewer plots) for animal ecologists to undertake for most groups.

The sampling requirements become more modest if significance levels are relaxed. For example, setting β = α = 0.10 reduced the sampling requirements in table by, on' average, 20 percent The main utility of these results is to provide a reference for animal ecologists to consult when planning monitoring activities or assessing the effectiveness of existing programs.

Note that stringent β and α levels (0.05) were used to

generate these results. Less stringent levels may well be more appropriate in a monitoring context.

A caveat is that these recommendations are based on the assumption that trends in populations are fixed and linear. This is appropriate in certain situations, such as declining endangered species or increasing introduced species, whose populations often follow deterministic trends.

However, most populations monitored follow an irregular trajectory. Furthermore, trends in a particular local population probably represent a random sample of a spatially variable, regional population trend. The simulation software described can accommodate random trend variation among plots or sites if estimates of its magnitude are available.

SETTING OBJECTIVES FOR A MONITORING PROGRAMME

It is important to emphasize that conclusions drawn from these analyses are contingent on the initial statement of a monitoring program's objectives. Power estimates are influenced by many factors controlled by researchers, such as duration and interval of monitoring, count means and variances, and number of sites and counts made per season.

Several other, somewhat arbitrary factors also exert an important influence on power estimates. These include trend strength (effect size), significance level (type I error rate), and the number of tails to use in statistical tests.

It is therefore critical that animal ecologists establish explicit and well-reasoned monitoring objectives before the initiation of any monitoring programme. These goals should address what magnitude of change in the population index is sought for detection, what probability of false detections will be tolerated, and what frequency of true declines can go undetected. An initial statement of objectives is important because subsequent efforts to judge the success or failure of a monitoring programme are made in terms of those objectives.

Identifying change in local populations is fraught with difficulties. Dubious population indices, bias in selection of survey sites, and weak design of monitoring programs can undermine trend detection. The practice of assessing population change in animal ecology could therefore be improved substantially.

First, one should not blindly assume that any readily measured population index can serve as a valid proxy for estimating actual abundance. As an alternative, performing simple pilot studies to ascertain the basic relationship between the index used and actual abundance will give animal ecologists much insight.

Such pilot studies can indicate whether the index used might yield misleading results, how it might be modified, and how it could potentially compromise trend detection. Second, animal ecologists must be aware of the potential pitfalls of nonrandom schemes for selecting sites for monitoring. A major challenge is to devise sampling methods that permit unbiased and statistically powerful surveys to be made in a logistically feasible manner.

Finally, conducting power analyses during the pilot phase of a monitoring programme is critical because it permits an assessment of a program's potential for meeting its stated goals while the opportunity for altering the program's structure is still available. The simulation method outlined and the summary of taxon-specific index variabilities can provide animal ecologists just such an option.

Successful monitoring of populations is based on making the best choices among sampling designs that yield precise estimates of a population index, statistical power considerations (trend strength, sample size, index variability, β, and α), and the statistical method used to analyse a count series. The primary consequence of failing to make the best choices and thereby improve methods for identifying population change in animal ecology will be a chronic failure to detect population change.

Unfortunately, these errors will often be misinterpreted as reflecting population stability, lack of treatment effect, or ineffectiveness of management. Neither the science of animal ecology nor the wild resources under our surveillance should be expected to bear the consequences of these errors.

Chapter 7

Population Viability Analysis

The biological diversity of the earth is threatened by the burgeoning human population. To prevent extinctions of species, conservationists must manage many populations in isolated habitat parcels that are smaller than desirable. An example is maintaining large-bodied predator populations in isolated, limitedarea nature reserves.

A population has been defined as "a group of individuals of the same species occupying a defined area at the same time". The viability of a population is the probability that the population will persist for some specified time. Two procedures are commonly used for evaluating the viability of a population. Population viability analysis (PVA) is the method of estimating the probability that a population of a specified size will persist for a specified length of time.

The minimum viable population (MVP) is the smallest population size that will persist some specified length of time with a specified probability. In the first case, the probability of extinction is estimated, whereas in the second, the number of animals that is needed in the population to meet a specified probability of persistence is estimated.

For a population that is expected to go extinct, the time to extinction is the expected time the population will persist. Both PVA and MVP require a time horizon: a specified but arbitrary time to which the probability of extinction pertains.

Definitions and criteria for viability, persistence, and extinction are arbitrary, such as a 95 percent probability of a population persisting for at least 100 years. Mace and Lande discussed criteria for extinction. Ginzburg et al. suggested the phrase "quasi-extinction risk" as the probability of a population dropping below some critical threshold, a concept also promoted by Ludwig and Dennis et al. Schneider and Yodzis used the term quasi-extinction to mean a population drop such that only 20 females remain.

The usual approach for estimating persistence is to develop a probability distribution for the number of years before the model "goes extinct," or falls below a specified threshold. The percentage of the area under this distribution in which the population persists beyond a specified time period is taken as an estimate of the probability of persistence.

To obtain MVP, probabilities of extinction are needed for various initial population sizes. The expected time to extinction is a misleading indicator of population viability because for small populations, the probability of extinction in the immediate future is high, even though the expected time until extinction may be quite large.

The skewness of the distribution of time until extinction thus makes the probability of extinction for a specified time interval a more realistic measure of population viability.

Simple stochastic models have yielded qualitative insights into population viability questions. But because population growth is generally considered to be nonlinear, with nonlinear dynamics making most stochastic models intractable for analysis, and because catastrophes and their distribution pose even more difficult statistical problems, analytical methods are generally inadequate to compute these probabilities.

Therefore, computer simulation is commonly used to produce numerical estimates for persistence or MVP. Analytical models lead to greater insights given the simplifying assumptions used to develop the model. However, the simplicity of analytical models precludes their use in real analyses because of the omission of important processes governing population change such as age structure and periodic breeding. Lack of data suggests the use of simple models, but lack of data really means lack of information.

Lack of information suggests that no valid estimates of population persistence are possible because there is no reason to believe that unstudied populations are inherently simpler (and thus justify simple analytical models) than well-studied populations for which the inadequacy of simple analytical models is obvious.

The most thorough recent review of the PVA literature was provided by Boyce. Shaffer, Soulé, Nunney and Campbell, and Remmert provided a historical perspective of how the field developed. Specifically, statistical methods to estimate the variance components needed to develop a PVA, the need to incorporate individual heterogeneity into a PVA, and the need to incorporate the sampling variance of parameter estimates used in a PVA are discussed.

QUALITATIVE OBSERVATIONS ABOUT POPULATION PERSISTENCE

Qualitatively, population biologists know a considerable amount about what allows populations to persist. Some generalities about population persistence are as follows:

- Connected habitats are better than disjointed habitats.
- Suitable habitats in close proximity to one another are better than widely separated habitats.
- Late stages of forest development are often better than younger stages.
- Larger habitat areas are better than smaller areas.
- Populations with higher reproductive rates are more secure than those with lower reproductive rates.
- Environmental conditions that reduce carrying capacity or increase variance in the growth rates of populations decrease persistence probabilities.

This list should be taken as a general set of principles, but you should recognize that exceptions occur often.

Typically, recovery plans for an endangered species try to create multiple populations of the species, so that a single catastrophe will not wipe out the entire species, and increase the size of each population so that genetic, demographic, and normal environmental uncertainties are less threatening.

However, Hess argued that connected populations can have lower viability over a narrow range in the presence of a fatal disease transmitted by contact. He demonstrated the possibilities with a model, but had no data to support his case. However, the point he made seems biologically sound, and the issue can be resolved only by optimizing persistence between these two opposing forces.

Spatial variation, that is, variation in habitat quality across the landscape, affects population persistence. Typically, extinction and metapopulation theories emphasize that stochastic fluctuations in local populations cause extinction and that local extinctions generate empty habitat patches that are then available for recolonization. Metapopulation persistence depends on the balance of extinction and colonization in a static environment. For many rare and declining species,

Thomas argued that extinction is usually the deterministic consequence of the local environment becoming unsuitable (through habitat loss or modification, introduction of a predator, etc.); that the local environment usually remains unsuitable following local extinction, so extinctions only rarely generate empty patches of suitable habitat; and that colonization usually follows improvement of the local environment for a particular species (or long-distance transfer by humans).

Thus persistence depends predominantly on whether organisms are able to track the shifting spatial mosaic of suitable environmental conditions or on maintenance of good conditions locally.

Foley used a model to agree that populations with higher reproductive rates are more persistent. However, mammals with larger body size can persist at lower densities and typically have lower annual and per capita reproductive rates. Predicted minimal density decreases as the "0.68 power of body mass, probably because of less variance in reproduction relative to life span in larger-bodied species.

The last item on the list—that environmental conditions that reduce carrying capacity or increase variance in the growth rates of populations decrease persistence probabilities—suggests that increased variation over time leads to lower persistence. One reason that increased temporal variation causes lowered persistence is that catastrophes such as hurricanes, fires, or floods are more likely to occur in systems with high temporal variation.

Populations in the wet tropics can apparently sustain themselves at densities much lower than those in temperate climates, probably because of less environmental variation. The distinction between a catastrophe and a large temporal variance component is arbitrary, and on a continuum. Furthermore, even predictable effects can have an impact. Beissinger modeled the effects of periodic environmental fluctuations on population viability of the snail kite (Rostrhamus sociabilis) and suggested that this source of variation is important in persistence.

CONTRADICTIONS

Few empirical data are available to support the generalities just mentioned, but exceptions exist. Berger addressed the issue of MVP by asking how long different-sized populations persist. He presented demographic and weather data spanning up to 70 years for 122 bighorn sheep (Ovis canadensis) populations in southwestern North America.

His analyses revealed that 100 percent of the populations with fewer than 50 individuals went extinct within 50 years, populations with more than 100 individuals persisted for up to 70 years, and the rapid loss of populations was not likely to be caused by food shortages, severe weather, predation, or interspecific competition.

Thus, 50 individuals, even in the short term of 50 years, are not a minimum viable population size for bighorn sheep. However, Krausman et al. questioned this result because they know of populations of 50 or less in Arizona that have persisted for more than 50 years.

Pimm et al. and Diamond and Pimm examined the risks of extinction of breeding land birds on 16 British islands in terms of population size and species attributes. Tracy and George extended the analysis to include attributes of the environment, as well as species characteristics, as potential determinants of the risk of extinction. Tracy and George concluded that the ability of current models to predict the risk of extinction of particular species on particular island is very limited. They suggested that models should include more specific information.

About the species and environment to develop useful predictions of extinction probabilities. Haila and Hanski criticized the data of Pimm et al. as not directly relating to extinctions because the small groups of birds breeding in any given year on single islands were not populations in a meaningful sense.

Although this criticism may be valid, most of the "populations" that conservation biologists study are questionable. Thus results of the analysis by Tracy and George do contribute useful information because the populations they studied are representative of populations to which PVA techniques are applied.

Specifically, small populations of small-bodied birds on oceanic islands (more isolated) are more likely to go extinct than are large populations of large-bodied birds on less isolated (channel) islands. However, interaction of body size with type of island (channel vs. oceanic) indicated that body size influences time to extinction differently depending on the type of island.

The results of Tracy and George support the general statements presented. As with all ecological generalities, exceptions quickly appear.

SOURCES OF VARIATION AFFECTING POPULATION PERSISTENCE

The persistence of a population depends on stochasticity, or variation. Sources of variation, and their magnitude, determine the probability of extinction, given the population growth mechanisms specific to the species. The total variance of a series of population measurements is a function of process variation (stochasticity in the population growth process) and sampling variation (stochasticity in measuring the size of the population). Process variation is a result of demographic, temporal and spatial (environmental), and individual (phenotypic and genotypic) variation.

NO VARIATION

Consider a population with no variation, one that qualifies for the simple, density-independent growth model N t + 1 = N t (1 + R), where N t is the population size at time t and R is the finite rate of change in the population. This model is deterministic, and hence, so is the population.

R e″ 0 guarantees that the population will persist, in contrast to R < 0, which guarantees that the population will go extinct (albeit in an infinite amount of time because a fraction of an animal is allowed in this model).R can be considered to be a function of birth and death rates, so that R = b ″ d defines the rate of change in the population as a function of birth rate (b) and death rate (d). When the birth rate exceeds or equals the death rate, the population will persist with probability 1 in this deterministic model.

STOCHASTIC VARIATION

Let us extend this naive model by making it stochastic. The parameter R to be a function of two random variables. At each time t, tochastically the number of animals to be added to the population by births and then the number to be removed by deaths.

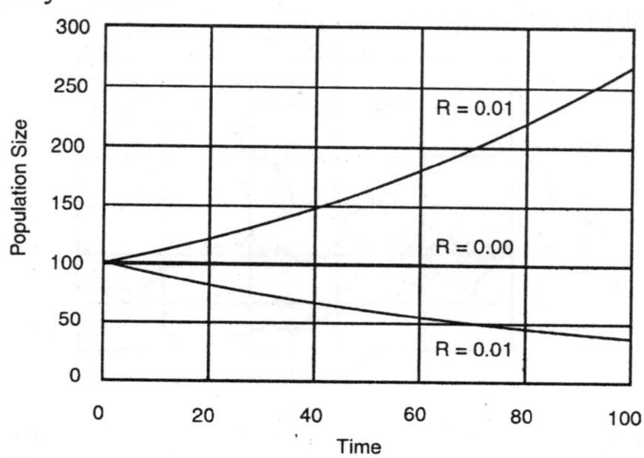

Fig. Deterministic Model of Population Growth.

For values of R e" 0, the population persists indefinitely. For values of R < 0, the population will eventually go extinct in that the number of animals will approach zero. Equals the death rate, say b = d = 0.5. That is, on average 50 percent of the N t animals would give birth to a single individual and provide additions to the population, and 50 percent of the N t animals would die and be removed from the population. Thus the population is expected to stay constant because the number of births equals the number of deaths.

A reasonable stochastic model for this process would be a binomial distribution. For the binomial model, you can think of flipping a coin twice for each animal. The first flip determines whether the animal gives birth to one new addition to the population in N t + 1 and the second flip determines whether the animal currently a member of N t remains in the population for another time interval, to be a member of N t + 1, or dies. If we start with N 0 = 100, what is the probability that the population will persist until t = 100?

You might be tempted to say the probability is 1 that the population will persist until t = 100 because the expected value of R is 0 given that the birth rate equals the death rate—that is, E(R) = 0, so that E(N t + 1) = E(N t). You would be wrong! Implementation of this model on a computer shows that the probability of persistence is 98.0 percent; that is, 2.0 percent of the time the population does not persist for 100 years without N t becoming 0 for some *t*.

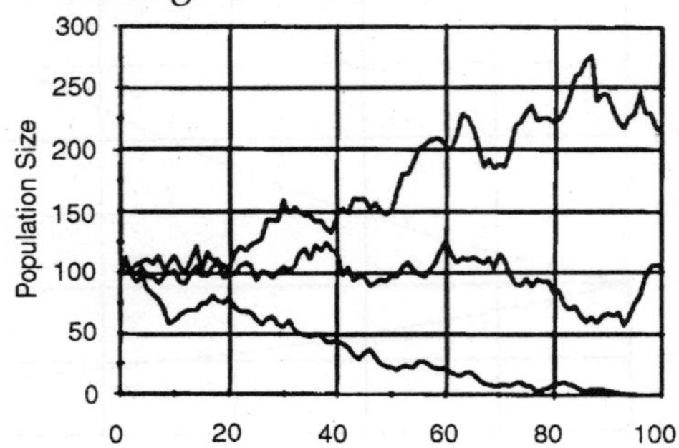

Fig. Three Examples of the outcome of the Population model with only Demographic Variation.

The smaller population goes extinct at time 93. Birth and death probabilities are both 0.5, making the expected value of R= 0.

These estimates were determined by running the population model 10,000 times and recording the number of times the simulated population went extinct before 100 years had elapsed. Lowering the initial population to N 0 = 20 results in persistence of only 53.2 percent of the populations, again based on 10,000 runs of the model. Setting N 0 = 500 improves the persistence rate to nearly 100 percent. Note that the persistence is not linear in terms of N 0. Initial population size has a major influence on persistence.

DEMOGRAPHIC VARIATION

Other considerations affect persistence. The value of R (the birth rate minus the death rate) is critical. R can be negative (death rate exceeds birth rate) and the population can still persist for 100 years, which may seem counterintuitive. Furthermore, R can be positive (birth rate exceeds death rate) and the population can still go extinct. For example, suppose R is increased to 0.02 by making the birth rate 0.51 and the death rate 0.49.

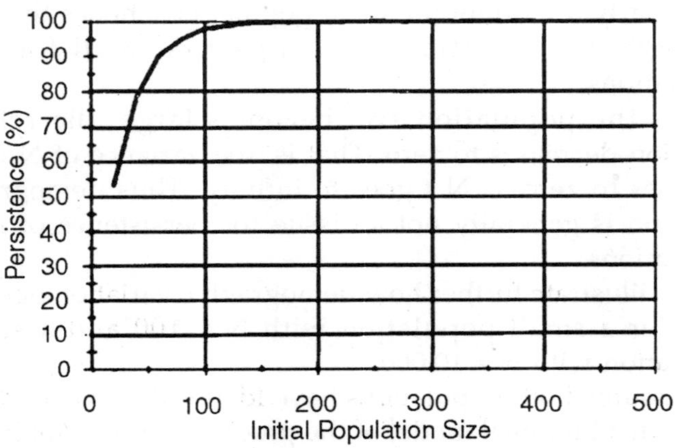

Fig. Persistence of a Population as a Function of Initial Population size (N 0) when only Demographic Variation is Incorporated into the model.

Birth and death probabilities are both 0.5, making the expected value of R = 0. The model was run 10,000 times to estimate the percentage of runs in which the population persisted until t = 100.

The persistence for N 0 = 20 increases to 84.3 percent from 53.2 percent for R = 0. Even though the population is expected to increase, stochasticity can still cause the population to go extinct.

The type of stochasticity illustrated by this model is known as demographic variation. This source of variation "penny-flipping variation" because the variation about the expected number of survivors parallels the variation about the observed number of heads from flipping coins.

To illustrate demographic variation, suppose the probability of survival of each individual in a population is 0.8. Then on average, 80 percent of the population will survive. However, random variation precludes exactly 80 percent surviving each time this survival rate is applied.

From purely bad luck on the part of the population, a much lower proportion may survive for a series of years, resulting in extinction. Because such bad luck is most likely to happen in small populations, this source of variation is particularly important for small populations, hence the name demographic variation. The impact is small for large populations.

As the population size becomes large, the relative variation decreases to zero. That is, the variance of N t + 1 / N t goes to zero as N t goes to infinity. Thus demographic variation is generally not an issue for persistence of larger populations.

To illustrate further how demographic variation operates, consider a small population with N = 100 and a second population with N = 10,000.

Assume both populations have identical survival rates of 0.8. With a binomial model of the process, the probability that only 75 percent or less of the small population survives is 0.1314 for the small population, but 3.194E " 34 for the larger population.

Thus the likelihood that up to 25 percent of the small population is lost in 1 year is much higher than for the large population.

TEMPORAL VARIATION

The variation of predicted population size increases with time. Some realizations of the stochastic process climb to very large population values after long time periods, whereas other realizations drop to zero and extinction. This result should be intuitive because as the model is projected further into the future, certainty about the projections decreases.

However, in contrast to population size, our certainty about the extinction probability increases as time increases to infinity. The probability of eventual extinction is always unity if extinction is possible. This is because the only absorbing state of the stochastic process is extinction; that is, the only population size at which there is no chance of change is zero.

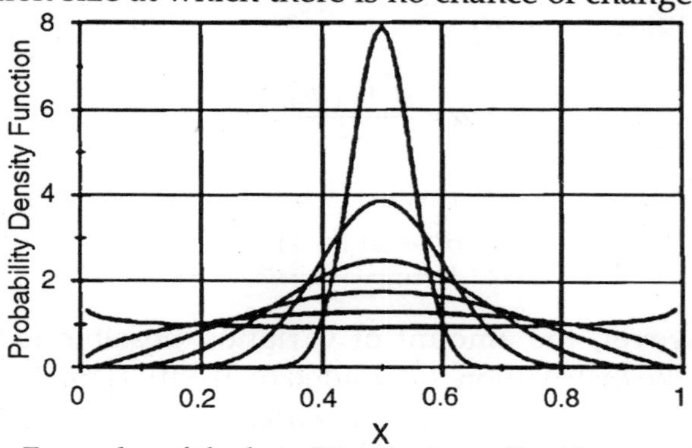

Fig. Examples of the beta Distribution, all with mean 0.5.

Another way to decrease persistence is to increase the stochasticity in the model. One way would be to introduce temporal variation by making b and d random variables. Such variation would be exemplified by weather in real populations. Some years, winters are mild and survival and reproduction are high.

Other years, winters are harsh and survival and reproduction are poor. To incorporate this phenomenon into our simple model, suppose that the mean birth and death rates are again 0.5, but the values of the birth rate and the death rate at a particular time t are selected from a statistical distribution, say a beta distribution. That is, each year, new values of b and d are selected from a beta distribution.

A beta distribution is bounded by the interval 0–1 and can take on a variety of shapes. For a mean of 0.5, the distribution is symmetric about the mean, but the amount of variation can be changed.

The beta distribution is described by two parameters, $\beta > 0$ and $\alpha > 0$. The standard deviations proceeding from the tallest curve to the lowest curve at $x = 0.5$ are 0.05 to 0.3 in increments of 0.05.

The mean of the distribution is given by $\alpha/(\alpha + \beta)$ and the variance as $\alpha\beta/[(\alpha + \beta)^2 (\alpha + \beta + 1)]$, with the mode $(\alpha - 1)/(\alpha + \beta - 2)$ (mode only for $\alpha \geq 1$). Most random number generation techniques for the beta distribution require you to specify values for α and β. For a given mean (μ) and variance (σ_2) or standard deviation (σ),

$$\alpha = \frac{\mu^2(1-\mu)}{\sigma^2} - \mu$$

and

$$\beta = \frac{\left[\sigma^2 + \mu(\mu-1)\right](\mu-1)}{\sigma^2}$$

However, the amount of variation possible is limited because the distribution is bounded on the [0, 1] interval. Thus for a mean of 0.5, the maximum variance approaches 0.25 as β and α approach zero.

The standard deviations of the birth and death rates over time affect persistence because these values determine the standard deviation of R. The smaller the standard deviations, the more the model approaches the demographic variation case, and thus, as N_t approaches infinity, the deterministic case. As the standard deviation increases, the more the variation in N_t, regardless of population size, and the less likely the population is to persist. Thus a standard deviation of 0.2 for both the birth and death rates results in only 28.5 percent persistence for N_0 = 100. Compare this to the 77.4 percent persistence achieved for a standard deviation of 0.1 or to the 98.0 percent persistence when no variation in birth and death rates occurred but demographic variation is still present.

This second source of variation in our simple model is temporal variation, that is variation in the parameters of the model across time. As the example shows, increasing temporal variation decreases persistence.

The simple model illustrated assumed that no correlation existed between the birth rate and the death rate, that is, that the two rates were independent. However, in real populations there is probably a high correlation between birth rates and death rates across years.

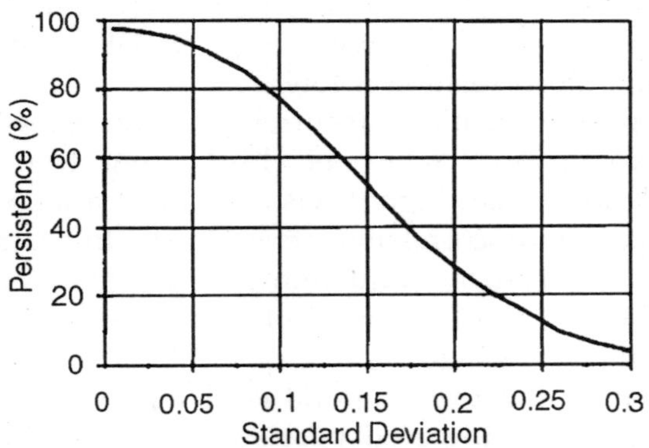

Fig. Persistence of a population of 100 animals at $t = 0$ to $t = 100$ years as a function of the standard deviation of birth (mean = 0.5) and death (mean = 0.5) rates (temporal variation). Demographic variation is still included in the model.

Good years with lots of high-quality resources available to the animals probably result in increased reproduction and survival, whereas bad years result in poor reproduction and high mortality. Including a negative covariance of birth and death rates (or a positive covariance between birth and survival rates) in the model results in an even bigger impact of temporal variation on persistence.

That is, the bad years are really bad because of both poor reproduction and high mortality, and the good years are really good. The net effect of this negative covariance of birth and death rates is to decrease persistence.

SPATIAL VARIATION

Spatial variation is the variation across the landscape that is normally associated with populations. Factors causing geographic variation include geologic differences that affect soil type, and thus habitat, and weather patterns (e.g., differences in rainfall across the landscape).

If the immigration and emigration rates are high across the landscape, so that subpopulations are depleted because of local conditions, high spatial variation can lead to higher persistence. This is because the probability of all the subpopulations of a population being affected simultaneously by some catastrophe is low when high spatial variation exists and spatial autocorrelation is low.

High positive spatial autocorrelation causes low levels of spatial variation, whereas high negative spatial autocorrelation causes high levels of spatial variation, as low levels of spatial autocorrelation generally do. In contrast, with low spatial variation (and hence high positive spatial autocorrelation), the likelihood of a bad year affecting the entire population is high.

Thus, in contrast to temporal variation, where increased variation leads to lowered persistence, increased spatial variation and low spatial autocorrelation lead to increased persistence, given that immigration and emigration are effectively mixing the subpopulations.

If immigration and emigration are negligible, then spatial variation divides the population into smaller subpopulations, which are more likely to suffer extinction from the effect of demographic variation on small populations. The combination of temporal and spatial variation is called environmental variation. Both dictate the animal's environment, one in time, one in space.

INDIVIDUAL VARIATION

All the models examined so far assume that each animal

in the population has exactly the same chance of survival and reproducing, even though these rates change with time. What happens if each animal in the population has a different rate of survival and reproduction? Differences between the individuals in the population are called individual heterogeneity, and this creates individual variation.

Many studies have demonstrated individual heterogeneity of individual survival and reproductions; for example, Clutton-Brock et al. (1982) demonstrated that lifetime reproductive success of female red deer (Cervus elaphus) varied from 0 to 13 calves reared per female.

Differences in the frequency of calf mortality between mothers accounted for a larger proportion of variance in success than differences in fecundity. Bartmann et al. demonstrated that overwinter survival of mule deer fawns was a function of the fawn's weight at the start of the winter, with larger fawns showing better survival.

Individual variation is caused by genetic variation, that is, differences between individuals because of their genome. Individual heterogeneity is the basis of natural selection; that is, differences between animals is what allows natural selection to operate. However, phenotypic variation is also possible, where individual heterogeneity is not a result of genetic variation.

Animals that endure poor nutrition during their early development may never be as healthy and robust as animals that are on a higher nutritional plane, even though both are genetically identical. Animals with access to more and better resources have higher reproductive rates, as in the red deer studied by Clutton-Brock et al. Thus individual heterogeneity may result from both genetic and phenotypic variation. Lomnickideveloped models of resource partitioning that result in phenotypic variation of individuals. Another example of individual heterogeneity in reproduction was provided by Burnham et al. in northern spotted owls (Strix occidentalis caurina).

In the case of northern spotted owls, repeated observations of reproduction across numerous individuals were used to estimate individual variation with analysis of variance procedures. The age of the female produced individual heterogeneity. This study also demonstrated temporal and spatial variation in owl fecundity rates.

Undoubtedly, natural selection plays a role in the genetic variation left in a declining population. Most populations for which we are concerned about extinction probabilities have suffered a serious decline in numbers. The genotypes remaining after a severe decline are unlikely to be a random sample of the original population.

The genotypes persisting through a decline are the "survivors," and would have a much better chance of persisting than would a random sample from the population before the decline. Of course, this argument assumes that the processes causing the decline remain in effect, so that the same natural selection forces continue to operate.

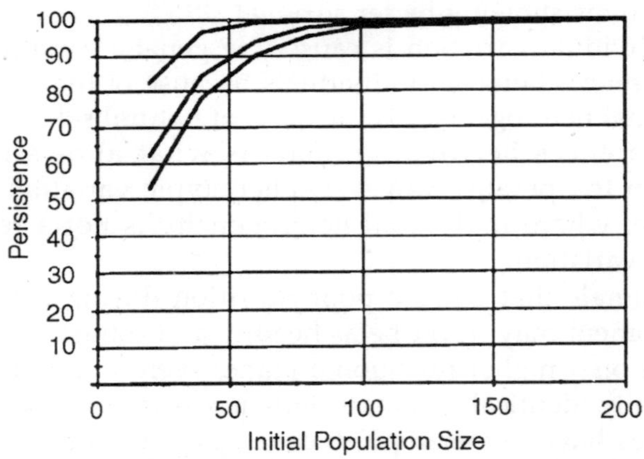

Fig. Effect of individual Variation on Population Persistence.

The three lines from top to bottom have standard deviations of 0.1, 0.05, and 0.01 for the birth and death rates.

Instead of each animal having exactly a birth rate of 0.5 and a death rate of 0.5, let's select these values from beta distributions with a mean of 0.5. The birth and death rates assigned to an animal remain for its lifetime. As new animals are added to the population, they likewise are assigned lifetime birth and death rates. How does persistence of this new model compare with the results from the demographic model? The reason that increased individual heterogeneity increases population persistence is that increased variation results in more chance that a few animals have exceptionally high reproductive potential and high survival. Therefore, these animals are unlikely to suffer mortality and be removed from the population and can be relied on to contribute new births each year.

The population may remain small but will not go extinct as often. Individual heterogeneity has seldom, if ever, been included in a population viability analysis, except as genetic variation. Yet as this simple example shows, individual heterogeneity not a function of genetic variation is a very important element in maintaining viability.

PROCESS VARIATION

The combined effect of demographic, temporal, spatial, and individual variation is called process varisation. That is, each of these sources of variation affects population processes. Process variation is used as a general term for the inherent stochasticity of changes in the population level. Process variation is in contrast to sampling variation, which is the variation contributed when biologists attempt to measure population processes.

That is, researchers are unable to measure the exact survival rate of a population. Rather, they observe realizations of the process, but not the exact value. Even if the fate of every animal in the population is observed, the resulting estimate of survival is only an estimate of the true but unknown population survival rate.

Methods of separating sampling variation from process variation are developed.Several lessons should be learned from this simple exercise. Persistence is a stochastic phenomenon. Even though the expected outcome for a particular model is to persist, random variation prevents this outcome from always occurring.

Small populations are much more likely to go extinct than larger populations because of demographic variation.

Increased temporal variation results in decreased persistence. Increased individual variation results in increased persistence.

COMPONENTS OF A PVA

Many factors affect the persistence of a population. What components are needed to provide estimates of the probability that a population will go extinct, and what are the tradeoffs if not all these components are available?

- A basic population model is needed. A recognized mechanism of population regulation, density dependence, should be incorporated because no population can grow indefinitely. "Of course, exponential growth models are strictly unrealistic on time scales necessary to explore extinction probabilities" (Boyce 1992:489).
  The population cannot be allowed to grow indefinitely, or persistence will be overestimated. Furthermore, the shape of the relationship between density and survival and reproduction can affect persistence, and density dependence cannot be neglected for moderate or large populations. Density dependence can provide a stabilizing influence that increases persistence in small populations.
- Demographic variation must be incorporated in this basic model. Otherwise, estimates of persistence will be too high because the effect of demographic variation for small populations is not included in the model.
- Temporal variation must be included for the parameters of the model, including some probability of a natural catastrophe. Examples of catastrophes (for some species) are fires (e.g., Yellowstone National Park, USA, 1988), hurricanes, typhoons, earthquakes, and extreme drought or rainfall resulting in flooding. Catastrophes must be rare, or else the variation would be considered part of the normal temporal variation. However, the covariance of the parameters is also important.

Good years for survival are probably also good years for reproduction. Likewise, bad years for reproduction may also lead to increased mortality. The impact of this correlation of reproduction and survival can drastically affect results.

For example, the model of Stacey and Taper (1992) of acorn woodpecker population dynamics performs very differently depending on whether adult survival, juvenile survival, and reproduction are bootstrapped as a triplet or given as individual rates across the 10-year period. If the positive correlation of the survival rates and reproduction is included in the model, estimated persistence is improved.

- Spatial variation in the parameters of the model must be incorporated if the population is spatially segregated. If spatial attributes are to be modeled, then immigration and emigration parameters must be estimated, as well as dispersal distances.

  The difficulty of estimating spatial variation is that the covariance of the parameters must be estimated as a function of distance; that is, what is the covariance of adult survival of two subpopulations as a function of distance?
- Individual heterogeneity must be included in the model or the estimates of persistence will be too low. Individual heterogeneity requires that the basic model be extended to an individual-based model. As the variance of individual parameters increases in the basic model, the persistence time increases.

  Thus, instead of just knowing estimates of the parameters of our basic model, we also need to know the statistical distributions of these parameters across individuals. This source of variation is not mentioned in discussions of population viability analysis such as Boyce, Remmert, Hunter, Meffe and Carroll, or Shaffer.
- For short-term projects, the sources of variation just mentioned may be adequate. However, if time

periods of more than a few generations are projected, then genetic variation should be considered. The population to change as selection takes place. Even if no selection is operating, genetic drift is expected for small population sizes.

However, the importance of genetic effects is still an issue in question; see Joopouborg and Van Groenendael. Lande suggested that demographic variation or genetic effects can be lethal to a small population.

- For long-term persistence, we must be willing to assume that the system will not change, that is, that the levels of stochasticity will not change through time, the species will not evolve through selection, and the supporting capacity of the environment (the species habitat) remains static.

  We must assume that natural processes such as long-term succession and climatic change do not affect persistence and that human activity will cease (given that humans have been responsible for most recent extinctions). To believe the results, we have to assume that the model and all its parameters stay the same across inordinately long time periods.

After examining this list, I am sure you will agree with Boyce: "Collecting sufficient data to derive reliable estimates for all the parameters necessary to determine MVP is simply not practical in most cases." Of course, limitations of the data seldom slow down modelers of population dynamics. Furthermore, managers are forced to make decisions, so modelers attempt to make reasonable guesses.

DIRECT ESTIMATION OF VARIANCE COMPONENTS

The implication of the list of requirements in the previous section is that population parameters or their distributions are known without error; that is, exact parameter values are observed, not estimated. In reality, we may be fortunate and have a series of survival or reproduction estimates across time that provides information about the temporal variation of the process. However, the variance of this series is not the proper estimate of the temporal variation of the process.

This is because each of our estimates includes sampling variation; that is we have only an estimate of the true parameter, not its exact value. To properly estimate the temporal variation of the series, the sampling variance of the estimates must be removed.

In this section, a procedure to remove the sampling variance from a series of estimates to obtain an estimate of the underlying process variation (which might be temporal or spatial variation). The procedure is explained in Burnham et al.

Consider the example situation of estimating overwinter survival rates each year for 10 years from a deer population. Each year, the survival rate is different from the overall mean because of snow depth, cold weather, and other factors. Let the true but unknown overall mean be S. Then the survival rate for each year can be considered to be S plus some deviation attributable to temporal variation, with the expected value of the e i equal to zero:

Environmental Variation

	Mean	*Year I*	*Year I*
1	S	$S + e_1$	S_1
2	S	$S + e_2$	S_2
3	S	$S + e_3$	S_3
4	S	$S + e_4$	S_4
5	S	$S + e_5$	S_5
6	S	$S + e_6$	S_6
7	S	$S + e_7$	S_7
8	S	$S + e_8$	S_8
9	S	$S + e_9$	S_9
10	S	$S + e_{10}$	S_{10}
Mean	S	$\bar{S}$	$\bar{S}$

The true population mean *S* is computed as $\bar{S}$:

$$\bar{S} = \frac{\sum_{i=1}^{10} S_i}{10}$$

with the variance of the *Si* is computed as.

$$\sigma^2 = \frac{\sum_{i=1}^{10} (S_i - \bar{S})^2}{10}$$

Where the random variables e_i are selected from a distribution with mean 0 and variance σ^2. In reality, we are never able to observe the annual rates because of sampling variation or demographic variation. For example, even if we observed all the members of a population, we would still not be able to say the observed survival rate was S_i because of demographic variation.

Consider flipping 10 coins. We know that the true probability of a head is 0.5, but we will not always observe that value exactly. If you have 11 coins; the true value is not even in the set of possible estimates. The same process operates in a population as demographic variation. Even though the true probability of survival is 0.5, we would not necessarily see exactly half of the population survive on any given year.

Hence, what we actually observe are the quantities following:

Environmental Variation +Sampling Variation

I	*Mean*	*Truth Year i*	*Observed Year I*
1	S	$S + e_1 + f_1$	S_1
2	S	$S + e_2 + f_2$	S_2
3	S	$S + e_3 + f_3$	S_3
4	S	$S + e_4 + f_4$	S_4
5	S	$S + e_5 + f_5$	S_5
6	S	$S + e_6 + f_6$	S_6
7	S	$S + e_7 + f_7$	S_7
8	S	$S + e_8 + f_8$	S_8
9	S	$S + e_9 + f_9$	S_9
10	S	$S + e_{10} + f_{10}$	S_{10}
Mean	S	$\bar{S}$	$\bar{S}$

Where the e i are as before, but we also have additional variation from sampling variation, or demographic variation, or both, in the fi.

The usual approach to estimating sampling variance

separately from temporal variance is to take replicate observations within each year so that withincell replicates can be used to estimate the sampling variance, whereas the between-cell variance is used to estimate the environmental variation. Years are assumed to be a random effect, and mixed-model analysis of variance procedures are used.

This approach assumes that each cell has the same sampling variance. An example of the application of a random effects model is Koenig et al. They considered year effects, species effects, and individual tree effects on acorn production by oaks in central California.

Classic analysis of variance methodology assumes that the variance within cells is constant across a variety of treatment effects. This assumption is often not true; that is, the sampling variance of a binomial distribution is a function of the binomial probability. Thus, as the probability changes across cells, so does the variance.

Another common violation of this assumption is caused by the variable of interest being distributed log-normally, so that the coefficient of variation is constant across cells and the cell variance is a function of the cell mean. Furthermore, the empirical estimation of the variance from replicate measurements may not be the most efficient procedure.

Therefore, the remainder of this section describes methods that can be viewed as extensions of the usual variance component analysis based on replicate measurements within cells. We examine estimators for the situation in which the within-cell variance is estimated by an estimator other than the moment estimator based on replicate observations.

Assume that we can estimate the sampling variance for each year, given a value of S_i for the year. For example, an estimate of the sampling variation for a binomial is,

$$\mathrm{var}(S_i \mid S_i) = \frac{S_i(1 - S_i)}{n_i}$$

where *n* i is the number of animals monitored to see whether they survived. Then, can we estimate the variance term due to environmental variation, given that we have estimates of the sampling variance for each year? If we assume all the sampling variances are equal, the estimate of the overall mean is still just the mean of the 10 estimates: $\bar{S} = \frac{\sum_{i=1}^{10} S_i}{10}$ with the theoretical variance being $(\bar{S}) = \frac{\sigma^2 + E[\text{var}(S \mid S)]}{10}$ i.e., the total variance is the sum of the environmental variance plus the expected sampling variance. This total variance can be estimated as $\text{var}(S) = \frac{\sum_{i=1}^{10} (S_{i-}\bar{S})^2}{10(10-1)}$ We can estimate the expected sampling variance as the mean of the sampling variances so that the estimate of the environmental variance is obtained by solving for σ^2

$$\sigma^2 \frac{\sum_{i=1}^{10} (S_{i-}\bar{S})^2}{(10-1)} - \frac{\sum_{i=1}^{10} \text{var}(S_i \mid S_i)}{10}$$

However, sampling variances are usually not all equal, so we have to weight them to obtain an unbiased estimate of σ^2. The general theory says to use a weight, w_i $w_i = \frac{1}{\sigma^2 + \text{var}(S_i \mid S_i)}$ so that by replacing var(S_i $S_{i,}$ S_i) with its estimator vâr(S_i S_i), the estimator of the weighted mean is $\bar{S} = \frac{\sum_{i=1}^{10} W_i S_i}{\sum_{i=1}^{10} W_i}$ with theoretical variance (i.e., sum of the theoretical variances for each of the estimates) $\text{var}(\bar{S}) = \frac{1}{\sum_{i=1}^{10} W_i}$

and empirical variance estimator $\text{var}(\bar{S}) = \dfrac{\sum_{i=1}^{10} W_i (S_i - \bar{S})^2}{\left(\sum_{i=1}^{10} W_i\right)(10-1)}$

When the w_i are the true (but unknown) weights, we have

$\dfrac{\cdot 1}{\sum_{i=1}^{10} W_i} = \dfrac{\sum_{i=1}^{10} w_i (S_i - \bar{S})^2}{\left(\sum_{i=1}^{10} w_i\right)(10-1)}$ giving the following $1 = \dfrac{\sum_{i=1}^{10} w_i (S_i - \bar{S})^2}{10-1}$

Therefore, all we have to do is manipulate this equation with a value of σ^2 to obtain an estimator of σ^2.

To obtain a confidence interval on the estimator of σ^2, we can substitute the appropriate chi-square values in this relationship. To find the upper confidence interval value, $\sigma^2 U$

solve the equation $\dfrac{\sum_{i=1}^{10} w_i (S_i - \bar{S})^2}{10-1} = \dfrac{\chi^2_{10-1,\alpha_L}}{10-1}$ and for the lower confidence interval value, σ^2_L solve the equation

$$\frac{\sum_{i=1}^{10} w_i (S_i - \bar{S})^2}{10-1} = \frac{\chi^2_{10-1,\alpha_U}}{10-1}$$

As an example, consider the following fawn survival data from overwinter survival of mule deer fawns at the Little Hills Wildlife Area, west of Meeker, Colorado, USA.

year	*Collared*	*Lived*	*Estimated Survival*	*Estimated Variance*
1981	46	15	0.3260870	0.0047773
1982	114	38	0.3333333	0.0019493
1983	118	5	0.0423729	0.0003439
1984	106	19	0.1792453	0.0013879
1985	155	59	0.3806452	0.0015210
1986	161	61	0.3788820	0.0014617
1987	116	15	0.1293103	0.0009706

The survival rates are the number of collared animals that lived divided by the total number of collared animals. For example, S_{1981} = 15/46 = 0.326087 for 1981. The sampling variance associated with this estimate is computed as

$$\text{var}(S_{1981}) = \frac{S_{1981}(1 - S_{1981})}{46}$$

which equals 0.0047773.

The procedure demonstrated here is applicable to estimation of other sources of variation (e.g., spatial variation) and to variables other than survival rates, such as per capita reproduction. The method is more general than the usual analysis of variance procedures because each observation is not assumed to have the same variance, in contrast to analysis of variance, in which each cell is assumed to have the same within-cell variance.

INDIRECT ESTIMATION OF VARIANCE COMPONENTS

Individual heterogeneity occurs in both reproduction and survival. Estimation of individual variation in reproduction is an easier problem than estimation of individual variation in survival because some animals reproduce more than once, whereas they only die once. Bartmann et al. demonstrated that overwinter survival of mule deer fawns is related to their mass at the start of the winter. Thus one approach to modeling individual heterogeneity is to find a correlate of survival that can be measured and develop statistical models of the distribution of this correlate. Then, the distribution of the correlate can be sampled to obtain an estimate of survival for the individual. Lomnicki also suggests mass as an easily measured variable that relates to an animal's fitness.

To demonstrate this method, use a simplification of the logistic regression model of Bartmann et al.:

$$\log\left(\frac{S}{1-S}\right) = \beta_0 + \beta_1 \text{ mass}$$

Where survival (S) is predicted as a function of weight.

Weight of fawns at the start of winter was approximately normally distributed, with mean 32 kg and standard deviation 4.2. To simulate individual heterogeneity in overwinter fawn survival, values can be drawn from this normal distribution to generate survival estimates.

This model can be expanded to incorporate temporal variation (year effects), sex effects, and area effects, as described for mule deer fawns by Bartmann et al.. An example of modeling temporal variation in greater flamingos (*Phoenicopterus ruber roseus*) as a function of winter severity is provided by Cezilly et al..

The approach suggested here of modeling winter severity as a random variable and estimating survival as a function of this random variable is an alternative to the variance estimation procedures of the previous section. Both provide a mechanism for injecting variation into a population viability model. The main advantage of using weather data to drive the temporal variation of the model is that considerably more weather data are available than are biological data on survival or reproductive rates.

The major drawback of the indirect estimation approach proposed in this section is that sampling variation of the functional relationship is ignored in the simulation procedure.

That is, the logistic regression model includes sampling variation because its parameters are estimated from observed data. The parameter estimates of the logistic regression model include some unknown estimation error. Their direct use results in potentially biased estimates of persistence, depending on how much sampling error is present. Therefore, a "good" model relating the covariate to the biological process is needed.

BOOTSTRAP APPROACH

Stacey and Taper used a bootstrap procedure to incorporate temporal variation into a model of acorn woodpecker (*Melanerpes formicivorus*) population viability. They used estimates of adult and juvenile survival and reproductive rates resulting from a 10-year field study to estimate population persistence. To incorporate the temporal variation from the 10 years of estimates, they randomly selected with replacement one estimate from the observed values to provide an estimate in the model for a year.

This procedure is known in the statistical literature as a bootstrap sampling procedure. The technique is appealing because of its simplicity. However, for estimating population viability, a considerable problem is inherent in the procedure. That is, the estimates used for bootstrapping contain sampling variation and demographic variation, as well as the environmental variation that the modeler is attempting to incorporate.

To illustrate how demographic variation is included in the estimates, consider an example population of 10 animals with a constant survival rate of 0.55. Thus, the actual temporal variation is zero, yet a sequence of estimates of survival from this population suggests considerable variation. That is, the estimates of survival would have a variance of 0.55(1-0.55)/10 = 0.02475 if all 10 animals had a survival probability of 0.55. Furthermore, the only observed values of survival would be 0, 0.1, ..., 1.0.

However, if the size of the population were increased to 100, you would find that the variance of the sequence of estimates becomes 0.002475, a considerable decrease. Thus randomly sampling the estimates from a population of size 10 results in considerably more variation than from a population of 100. The demographic variation from the sampled population is incorporated into the persistence model if the bootstrap approach is used.

A similar example can be used to demonstrate that sampling variation is also inherent in bootstrapping from a sample of observed estimates. Suppose a sample of 10 radiocollared animals is used to estimate survival for a population of 100,000 animals (i.e., the finite sample correction term can be ignored). The sampling variation of the estimates would be S(1- S)/10, where S is the true survival rate for the population assuming all animals had the same survival rate.

Now if a sample of 100 radiocollared animals is taken,

the sampling variation reduces to S(1-S)/100. Thus randomly sampling estimates with a bootstrap procedure incorporates the sampling variation of the estimates into the persistence model.

BASIC POPULATION MODEL AND DENSITY DEPENDENCE

Leslie matrix models are commonly used as the modeling framework for population viability models. Density dependence must be incorporated into the model; that is, basic parameters must be a function of population size. Thus the resulting model is not a true Leslie matrix.

Each iteration of the calculation also requires a temporal variance component, and making the parameters of the Leslie matrix into random variables is the standard approach but eradicates the analytical results that normally are benefits of Leslie's creative work. If multiple patches are modeled, each patch requires a spatial variance component. Demographic variation can be built into the model. Still, the resulting model doesn't resemble the elegant matrix model that Leslie originally developed.

However, use of the Leslie matrix framework ignores individual heterogeneity, and thus is likely to underestimate persistence. Incorporation of individual heterogeneity requires an individual-based model and thus is conceptually different from the basic Leslie matrix approach. Individual-based models can be spatially explicit, providing another approach to incorporating spatial stochasticity into the model.

As suggested by Boyce, Stacey and Taper, and Burgman et al. density dependence is an important part of estimating a population's persistence. Lande demonstrates that the importance of environmental stochasticity and random catastrophes depends on the density-dependence mechanism operating in the population, based on the value of K carrying capacity.

However, how density dependence is incorporated into the model greatly affects the estimates of persistence. In persistence models, as a population declines, compensation for small population size takes the form of increased birth rates and decreased death rates (density dependence) and so is a significant factor in increasing population persistence.

Consider the model $N_{t+1} = N_t\,[1 + R\,(t)]$ Stacey and Taper (1992) tested two forms of density dependence with their data; the logistic form $R(t) = R_0\left(1 - \frac{N(t)}{K}\right)$ and the θ-logistic form

$$R(t) = R_0\left[1 - \left(\frac{N(t)}{K}\right)^{\theta}\right]$$

Expressed as a difference equation, the θ-logistic model would be

$$N_{t+1} = N_t\left\{1 + R_0\left[1 - \left(\frac{N(t)}{K}\right)^{\theta}\right]\right\}$$

For θ = 1, the two models are identical. Although Stacey and Taper's data precluded a significant test between these models, their data did show significant correlations between adult survival and population size, suggesting that density dependence was operating in the population.

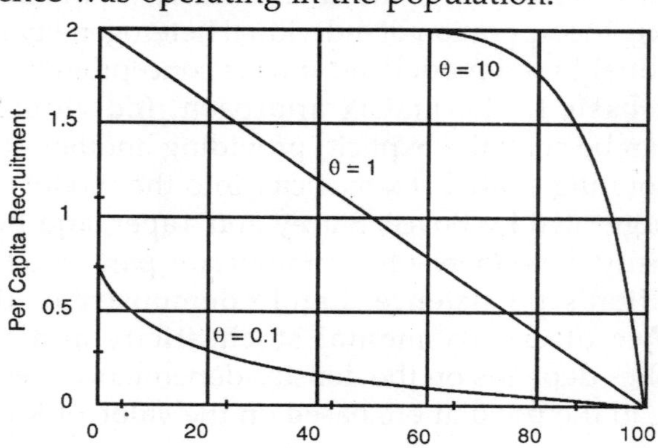

Fig. Three Examples of Possible Relationships of Recruitment per Individual to Population size (N_t).

The distinction between the two models can be very

important. In the first, the rate of change of the birth and death rates with population size is linear (i.e., the classic logistic population growth model). In the second, the change can be very nonlinear. The θ-logistic model can cause populations to be very persistent, or very extinction prone, depending on the shape of the function.

The curve for per capita recruitment with $\theta = 10$ results in a population with much greater persistence than the curve with $\theta = 0.1$ because as the population size becomes small, the $\theta = 10$ population is at peak reproduction for populations below 60, whereas peak reproduction is reached only at a population size of zero for the $\theta = 0.1$ population.

Typical sigmoid population growth demonstrated by the logistic curve results for $\theta = 1$. A steeper curve initially results for $\theta = 0.1$, whereas a flatter curve initially results for $\theta = 10$.

Burgman et al. and May and Oster summarize other functional relationships to incorporate density dependence. Possibilities, expressed as a difference equations, include those by Hassell, Hassell et al., and May:

$$N_{t+1} = \frac{\lambda N_t}{(1 + aN_t)^b}$$ by Moran and Ricker:

$$N_{t+1} = N_t \exp\left[r\left(1 - \frac{N_t}{K}\right)\right]$$

By Pennycuick et al., Usher, and Beddington, taken from May and Oster: by Beverton and Holt and Ricker:

$$N_{t+1} = \frac{1}{\rho + (k \mid N_t)}$$ and by Maynard-Smith and Slatkin:

$$N_{t+1} = \frac{R_0 N_t}{1 + (R_0 - 1)\left(\frac{N_t}{k}\right)^c}$$

Each of these models results in a different relationship between per capita recruitment and population size. Furthermore, these simple models can be applied to various segments of the life cycle, such as fecundity rates, neonatal survival, and adult survival, to achieve more realistic biological models. But the use of different models means that density dependence is implemented differently at a particular population level and population viability is affected. For example, Mills et al. reported widely differing estimates of population viability of grizzly bear (Ursus arctos horribilis) depending on which of four computer programs were used to compute the estimate.

Probably part of the discrepancy is in how density dependence was implemented in each of the programs, but different functions probably were used and these relationships probably were applied to differing segments of the life cycle. Unfortunately, distinguishing between these various models of density dependence with data is not practical because of the stochasticity (noise) in observed population levels, as Pascual et al. demonstrated by fitting a collection of models to Serengeti wildebeest (Connochaetes taurinus) data.

Fowler argues that both theory and empirical information support the conclusion that most density-dependent change occurs at high population levels (close to the carrying capacity) for species with life history strategies typical of large mammals, such as deer ($\theta > 1$). The reverse is true for species with life history strategies typical of insects and some fishes, with $\theta < 1$).

Note that explicit estimates of carrying capacity (*K*) and its variance are not needed to incorporate density dependence into a population model, although such an approach is possible. If the functional relationships between birth and death rates and population density are available, the carrying capacity is determined by these relationships.

Furthermore, if these relationships incorporate temporal and spatial variation, then the resulting model will have temporal and spatial variation in its carrying capacity, and thus stochastic density dependence.

Another example of how density dependence can operate in small populations is provided by the Allee effect: The per capita birth rate declines at low densities because, for example, of the increased difficulty of finding a mate. This is known as Allee-type behaviour (of the per capita birth rate), and its effect on the per capita population growth rate, $R(t)$, is called an Allee effect. In theory, a low-density equilibrium would be sustained in a deterministic equilibrium, where the birth rate equals the death rate. However, given stochasticity, the population could be driven below the low density equilibrium, and thus slide into extinction.

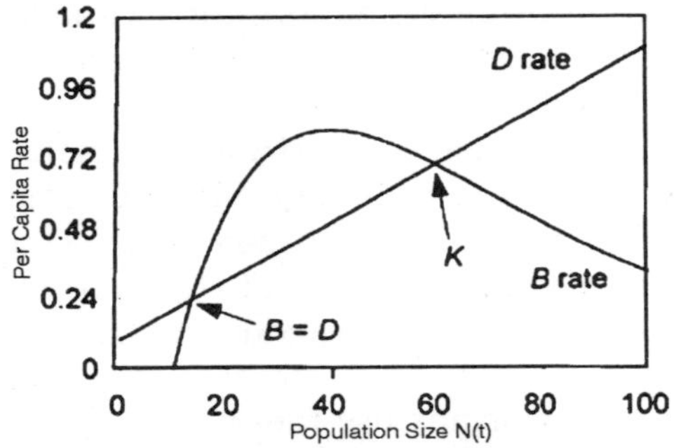

Fig. Example of how an Allee effect is Created by a Declining birth Rate at Low Densities.

Two equilibrium points exist, where the birth rate equals the death rate. The lower equilibrium may be stable in a deterministic system but could easily lead to extinction in a stochastic system.

A second example of incorporating density dependence into a population viability analysis is provided by Armbruster and Lande. They used estimates of life history parameters of elephant (Loxodonta africana) populations in a fluctuating environment from studies at Tsavo National Park, Kenya, to develop an age-structured, density-dependent model.

Density regulation was implemented by changes in the age of first reproduction and calving interval. They modeled environmental stochasticity with drought events affecting sexand age-specific survival.

INCORPORATION OF PARAMETER UNCERTAINTY INTO PERSISTENCE ESTIMATES

To remove the sampling variation from estimates of process variation. Unbiased estimates of process variances such as temporal and spatial variation can be achieved. The parameter estimates into the estimates of persistence and, in the process, provide an unbiased estimate of persistence given the population model.

Any model developed to estimate population persistence has several to many parameters that must be estimated from available data. Each of these estimates has an associated estimate of its precision in the form of a variance, assuming that statistically rigorous methods were used to estimate the parameter from data.

In addition, because some of the parameters may have been estimated from the same set(*s*) of data, some parameters in the model may have a nonzero covariance. Thus, the vector of parameter estimates θ used in the model to estimate persistence has the variance—covariance matrix to measure uncertainty.

Typically, statisticians use the delta method to estimate the variance of a function of parameters from a set of parameter estimates and their variance—covariance matrix. In the context of persistence, the sampling variance of the estimate of persistence (p^) would be estimated from the sampling variances of the parameters in the model as

$\text{var}(p) = \frac{\delta f(\theta)^T}{\delta\theta}\text{var}(\theta)\frac{\delta f(\theta)}{\delta\theta}$ where p^ = f (θ). That is, the function f represents the model used to estimate persistence. However, for realistically complex persistence models, the analytical calculation of partial derivatives needed in this formula is probably not feasible.

The lack of explicit analytical partial derivatives suggests that numerical methods be used. The most feasible, albeit numerically intensive, appears to be the parametric bootstrap approach (Efron and Tibshirani 1993; Urban Hjorth 1994). With a parametric bootstrap, a realization of the parameter estimates is generated based on their point estimates and sampling variance—covariance matrix using Monte Carlo methods.

A multivariate normal distribution probably is used as the parametric distribution describing the set of parameter estimates, although other distributions or combinations of distributions may be more realistic biologically. Using this set of simulated values in the persistence model, persistence is estimated. This step requires a large number of simulations to properly estimate persistence with little uncertainty; typically 10,000 simulations are conducted. Then, a new set of parameter values is generated and persistence again estimated.

This process is repeated for many sets of parameter estimates (at least 100, but more likely 1,000) to obtain a set of estimates of persistence. The variation of the resulting estimates of persistence is then a measure of uncertainty attributable to the variation of the parameter estimates as measured by their variance—covariance matrix. The process is diagrammed as

Parametric bootstrap loop (1,000 iterations): Select realization of parameter estimates monte carlo loop (10,000 iterations): Tabulate percentage of model runs resulting in persistence end monte carlo loop end parametric bootstrap loop

However, even more critical to our viability analysis is the fact that the mean of this set of 1,000 estimates of persistence is probably less than the estimate we obtained using our original point estimates of model parameters. More formally, the expected value of estimated persistence $[E(P)]$ is less than the value of persistence predicted by our model using the point estimates of its parameters, $E(p) < f\,[E(\theta)]$ an example of Jensen's inequality.

This difference is caused by large probabilities of early extinction for certain parameter sets that are likely, given their sampling variation (Ludwig 1996a). Therefore, to estimate persistence, the mean of the bootstrap estimates of persistence should be used, not the estimate of persistence obtained by plugging our parameter estimates directly into our population model.

Confidence intervals on persistence could be constructed using the usual ±2 SE procedure based on the set of 1,000 estimates. This confidence interval represents the variation attributable to the uncertainty of the parameter estimates used in the model. Uncertainty about the model is not included in this confidence interval because the model is assumed to be known. However, a better confidence interval will probably be achieved by sorting the 1,000 values into ascending order and using the 25th and 975th values as a 95 percent confidence interval. This procedure accounts for the likely asymmetric distribution of the estimates of persistence.

The real problem with PVA is not the model, but obtaining the data to drive the model. Much of the published work on PVA ignores this essential point. For example, Mangel and Tier simplify the process to the point that they miss major issues concerning data reliability and quality of the product (estimates of persistence). Their four "facts" are as follows:

- "A population can grow, on average, exponentially and without bound and still not persist." This is because of catastrophes that will bring even a thriving population to zero.
- "There is a simple and direct method for the computation of persistence times that virtually all biologists can use." They suggest a simple model with one age class and a population ceiling that the population cannot exceed, but the ceiling does not cause density dependence effects of growth parameters. Their approach to estimating persistence is likely to underestimate persistence if the ceiling is set too low because the population can never grow away from the absorbing state of extinction.
- "The shoulder of the MacArthur—Wilson model occurs with other models as well, but disappears when catastrophes are included." They suggest a slow, steady rise in persistence times as the population ceiling is increased.
- "Extinction times are approximately exponentially distributed and this means that extinctions are likely." Thus, they conclude the most likely value of a population is zero, or the mode of an exponential distribution. This result is obtained because of the simplistic assumptions they have used.

Another misguided example is Tomiuk and Loeschcke.

Their mathematics cover up the real problem of obtaining realistic estimates of the parameter values to use in the models. Their model emphasizes demographic variation and ignores the bigger issues of temporal variation and individual heterogeneity.

A common problem with PVA is that the sampling variation of the parameter estimates is ignored. In both cases, estimates of persistence are too pessimistic because the sampling variation of the population parameters is included in the population model as if it were temporal variation. Furthermore, individual heterogeneity was left out of the model, further biasing the estimates of persistence too low.

"Most PVAs have ignored fundamentals of ecology such as habitat, focusing instead on genetics or stochastic demography". For small populations (less than 50) of endangered species, such a strategy may be justified, particularly for short-term predictions.

But incorporating only demographic variation results in overestimates of persistence because temporal variation has been ignored. On the other hand, the remaining survivors of an endangered species may be the individuals with high survival and reproductive rates, so the lack of individual heterogeneity may underestimate persistence.

The studies just mentioned should not lead the reader to believe that useful attempts to estimate persistence do not exist. Schneider and Yodzis developed a model of marten (Martes americana) population dynamics that incorporated the behaviour and physiology of individual martens, spatial dynamics, and demographic and environmental stochasticity.

Undoubtedly some readers would quibble with some of the assumptions and data used to build the model, but a realistic model with some of the inputs guessed (and clearly stated to be such) is a much more reasonable approach than a simplistic model that ignores important processes affecting persistence. Furthermore, such realistic models identify data needs that can be addressed with time, even though the actual estimate of persistence is of questionable value.

The alternative of using simplistic and naive models

leads to invalid estimates and little progress in improving the situation, with a rapid loss of credibility by the field of conservation biology.

Murphy et al.proposed two different types of PVA. For organisms with low population densities that are restricted to small geographic ranges (typical vertebrate endangered species), genetic and demographic factors should be stressed.

For smaller organisms such as most endangered invertebrates, environmental uncertainty and catastrophic factors should be stressed because these organisms are generally restricted to a few small habitat patches, but are capable of reaching large population sizes within these patches.

Nunney and Campbell noted that demographic models and genetic models both have resulted in similar estimates of minimum viable population size, but that the ideal spatial arrangement of reserves remains an issue.

Lande suggests that genetic mutations may affect fitness, so ignoring genetic effects results in underestimates of viability. Mutation can affect the persistence of small populations by causing inbreeding depression, by maintaining potentially adaptive genetic variation in quantitative characters, and through the erosion of fitness by accumulation of mildly detrimental mutations.

Populations of 5,000 or more are required to maintain evolutionary viability. Theoretical results suggest that the risk of extinction caused by the fixation of mildly detrimental mutations may be comparable in importance to environmental stochasticity and could substantially decrease the longterm viability of populations with effective sizes as large as a few thousand. If these results are correct, determining minimum viable population numbers for most endangered species is an exercise in futility because almost all of these populations are already below 5,000.

Conservation biologists would like to have rules to evaluate persistence, such as the magical Franklin—Soulé number of 500 that is the effective population size (N_e) to maintain genetic variability in quantitative characters.

Unfortunately, these rules lack the realism to be useful.

The Franklin—Soulé number was derived from simple genetic models and hence lacks the essential features of a PVA model discussed here. Attempts with simplistic models such as Mangel and Tier and Tomiuk and Loeschcke also do not provide defensible results because of the lack of attention to the biology of the species and the stochastic environment in which the population exists.

Until conservation biologists do good experimental studies to evaluate population persistence empirically, the usefulness of the rules and simplistic models suggested in some of the literature. PVA can be viewed as a heuristic tool to explore the dynamics of an endangered population but not as a predictive tool. PVA could be used to identify variables to which the population may be sensitive and to investigate the relative benefits of alternative kinds of management.

Some readers will argue that in this context, the absolute reliability of the model estimates of extinction probability, or time to extinction, matters much less than the extent to which risk is affected by different demographic and environmental variables. PVA so strongly depend on which sources of variation are included in the model and their relative magnitudes.

For example, the importance of demographic variation is stressed in PVA because it happens to be the simplest source of variation to model, generally requiring only the assumption of binomial variation.

Temporal variation has received less emphasis because it is a more difficult to obtain estimates of the temporal variance of population parameters. Individual heterogeneity has received no attention because this source of variation is by far the most difficult source to quantify, particularly for survival rates.

The only way to draw valid inferences about the importance of various sources of stochasticity affecting a population is to have reasonably good estimates of these parameters. Simplistic PVA models based on few or no data lead to simplistic and unreliable answers.

Without data, why would you expect anything else?Until rigorous experimental work can be conducted, conservation biologists should borrow information from game species, where long-term studies have been done that will provide estimates of temporal and spatial variation and individual heterogeneity.

Rules that predict temporal variation in survival as a function of weather, or individual variation in survival as a function of body characteristics, provide alternative sources of data.

For at least some game species, data exist to develop such rules. Furthermore, these kinds of data probably will never be available for many endangered species; the opportunity to collect such data was lost with the decline of the population to current (threatened) levels.

Taxonomically related species may provide information, although species in the same ecological guild may also provide information on temporal and spatial variation. Note the distinction between using estimates of the temporal and spatial variation and individual heterogeneity from a related species and using estimates of survival and recruitment from a surrogate species.

Estimates of survival and recruitment from a stable or increasing population would obviously be inappropriate for a species with a declining population. In summary, most estimates of population viability are nearly useless because one or more of the following mistakes or omissions are made in developing a model to estimate persistence. By listing omissions, this list suggests the essential ingredients to develop a useful PVA.

- Few or no data are available to estimate basic parameters in the population model, with almost all the parameter estimates just guesses. The resulting estimate of persistence is therefore strictly a guess. Lesson: To do a valid PVA, you must have data to build a realistic population model.
- The model ignores spatial variation, which increases population viability. As suggested by Stacey and

Taper (1992), immigration can occasionally rescue a population from extinction.

Lesson: If the population is widely distributed geographically, incorporate spatial variation.

- The model uses estimates of temporal variation that are at best poor guesses. This statement assumes that the modeler understood the difference between process variation and sampling variation. Often, sampling variation is assumed to substitute for process variation; the estimates of persistence are too pessimistic. Sampling variation has nothing to do with population persistence. Estimates of population parameters must not be treated as if they are the true parameter value.

Lesson: Obtain reliable estimates of temporal variation and don't confuse sampling variation and temporal variation.

- The model uses demographic variation as a substitute for temporal variation in the process and ignores true temporal variation.

Lesson: Incorporate both demographic and temporal variation into the PVA.

- The model ignores life-long individual heterogeneity that increases population viability and assumes that all individuals endure the same identical survival and reproduction parameters. Such a naive assumption results in population viability being underestimated.

Lesson: Individual heterogeneity must be incorporated into a PVA model if you don't want to underestimate viability.

- The model assumes that current conditions are not changing; that is, the stochastic processes included in the model are assumed constant for the indefinite future. Loss of habitat and other environmental changes that affect these stochastic processes are ignored. Thus, as discussed by Caswell (1989), the model probably is not useful in forecasting (i.e., predicting what will happen) but is useful in projecting (i.e., predicting what will happen if conditions do not change).

Lesson: Recognize that your model does not predict the future; it only projects what might happen if the system doesn't change (which is unlikely).

Before you use the estimates of persistence from any population viability analysis, compare the approach used to obtain the estimate with the necessary components discussed here.

If you discover omissions and errors in the approach used to obtain the estimate, recognize the worth (or lack thereof) of the estimate of persistence. Although the estimates of persistence obtained from a PVA may have little value, the process of formulating a model and identifying missing information (i.e., parameters that are poorly estimated) may still have value in developing measures to conserve the species in question.

Chapter 8

Mammalian Societies

Today, biologists interpret behaviour within a context fortified by theories of cognition, behavioral evolution, and games, and any or all of four processes may lead to cooperation: kin selection, reciprocity and byproduct mutualism, and even trait-group selection.

The processes that fashion societies are set within an ecological context, and a species' ecology can scarcely be interpreted without understanding its social life. As the specialties within whole-animal biology diversify and the once close-knit family of behavioral and ecological disciplines risks drifting apart, our purpose is to alert ecologists to the ethologist's tools for measuring social dynamics.

SOCIAL DYNAMICS

"If animals live together in groups their genes must get more benefit out of the association than they put in". What methods are available to measure the negotiations—the social dynamics—in this profit and loss account? We define a social dynamic simply as the change in social interaction or relationship under the influence of extrinsic or intrinsic factors. Our purpose here is to show how these changes and the factors influencing them may be measured and identified.

Likely candidates include the forces of ecological and demographic change, together with changes in the experiences and characters of group members. Ontogenic effects (individuals growing up and changing roles) might also affect the long-term social dynamics of a group that change the demography and hence the character of the society.

Effects on social dynamics may be erratically stochastic or predictably circadian, seasonal, annual, or of an even longer periodicity, largely following environmental rhythms. Predictable changes in social structure may also follow as a population or group progresses in a social succession toward carrying capacity after colonization or population crashes. Against this backdrop of almost continual flux, the study of social dynamics requires the measurement of changes in behavioral parameters.

These measures become the currency with which to assess predictions designed to test whether the forces of change have been isolated correctly. The concept of a group's social dynamic is a vital and often neglected foil to attempts to characterize a typical social structure. Therefore, the concept of social dynamics lies at the interface of sociobiology, ethology, and behavioral ecology and even includes aspects of complexity theory and emergent systems. This alone makes it a topic of dauntingly large scope.

Research on social behaviour commonly seeks a conclusion as to whether a particular type of social interaction maximizes fitness. However, to avoid the hazards of naive interpretation, one cannot draw such a conclusion without knowing the pattern of other interactions within which the behaviour in question is set. Behavioral ecologists may pick individuals for which they score an approximation of fitness against a continuum of strategies.

This approach is more hazardous as the web of social interactions in which individuals of a species are enmeshed becomes more complex. Occam's razor may suggest making the simplest explanation on the basis of what you observe, but in a social network the system is seldom simple, so it is prudent to make those observations thoroughly and in a wide context before that razor can be wielded confidently.

That the social dynamics of a species are both determinants and consequences of its ecology may be clear. To a field ecologist seeking to understand any part of this loop, it may be much less obvious how to characterize a social system in replicable, enduring, and quantitative terms as a basis for modern analysis. Historically, ethology pursued its own agenda—often with captive primates—of characterizing societies by observing behavioral interactions and directionality of behaviors within groups.

Classic ethologists were careful to record the detail of behaviour with a view to allowing comparisons between studies and between species; although modern comparative methods have brought elegance to the task of making comparisons, modern fashions have drifted away from assembling the data classic ethologists so carefully gleaned.

When we launched into this task, no text existed to bring together the wisdom of a receding era of ethology and the innovations promised by avant garde methods; as we finished, Lehner's comprehensive second edition Handbook of Ethological Methods was published. Lehner's updated text is set to become the benchmark.

Our treatment has five sections. First we mention three reasons why documenting social dynamics is important. Second, we tackle the question of how to describe social dynamics in ways that provide a framework within which to compare studies; in this, we follow Hinde in describing a hierarchical approach to the study of social dynamics.

We show that single behavioral interactions are the fundamental unit of social structure, and that it is the changing nature of repeated interactions—relationships—that gives social structure its potentially dynamic component. Third, we follow this framework to discuss how behavioral parameters can be described, classified, and recorded during social interactions.

In this we explore interactions during grooming and dominance. In the fourth section we explore methods used for gathering data on social interaction and then, in the fifth section, we use example data to illustrate analytical techniques for elucidating relationships, and how patterns of relationships can be combined to reveal social networks and social structure.

Because mammalian societies are ethologically complex and because our own experience is largely with mammals, we draw our examples from that class. Furthermore, we often use examples from our own work, not because it deserves mention more than any other, but because we know most well the lessons learned and pitfalls encountered therein; our intention is merely to illustrate points, not to review them compendiously.

Throughout, we focus on common mistakes in approach, methodological conflicts, and the use of new technology to solve problems (and how it creates some new ones). We have doubtless fallen short of our own prescriptions on many previous occasions, and are aware that many methods outlined here could be improved further.

WHY STUDY SOCIAL DYNAMICS?

Social dynamics merit study for three major reasons. First, the changing relationships between individuals are the building blocks from which dynamic social structures are assembled. Understanding their emergent properties, such as dominance hierarchies and social competition, is essential to tackling fundamental questions about the evolution of sociality.

Second, because social dynamics are the product of interaction between individuals' ecology and behaviour, they are relevant to predicting and managing the consequences of many human interventions for conservation or management. Third, an important motivation for understanding nonhuman societies is the light this throws on human behaviour. Each of these three topics is vast, but we fleetingly mention them in turn.

EVOLUTION OF SOCIALITY

The diverse relationships of individuals in a social network interact to create complex emergent patterns. These patterns, like the vortex that appears in an emptying sink, is not contained in the structure of a single component. Because a society represents a whole with properties different from those of its component parts, the ultimate consequences of social interactions may be remote from an observed action.

This is a fact that evolution by natural selection can take in its stride but that we, as primarily linear cause-and-effect thinkers, may find hard to accommodate. It may be clear that a lion killing a zebra is behaving adaptively, but less clear whether it is adaptive when the same lion prevents a conspecific from feeding at the kill. The immediate effect is that the first lion may have more food to eat, but the ultimate effects reverberate through a stochastically unpredictable system of long-term consequences among the whole pride.

Denied food or coalitionary aid by an ally of the snubbed individual at a later date, the fitness consequences for the originally possessive lion may be far from advantageous. An understanding of social dynamics offers insight into the adaptation of individual responses evolved from selection operating on them from the level of emergent systems.

CONSERVATION APPLICATIONS

Many problems in wildlife conservation and management involve humans causing changes to animal populations or their environment. In applied work, an understanding of the dynamics of a social system is a prerequisite to predicting the effect of human activities on, for example, spatial organization, population dynamics, and dispersal.

For example, attempts to control the transmission of bovine tuberculosis by killing badgers, a reservoir of the disease, clearly disrupts the society of survivors. The effects of such perturbation on social dynamics may alter the transmission of the disease, plausibly for the worse. A similar case may be argued regarding rabies control.

Translocation of elephants without regard for the social structure that provides adolescent discipline has led to problem animals in some African parks. Tuyttens and Macdonald (in press) review some consequences of behavioral disruption for wildlife management. Population control has been shown to affect the rate and pattern of dispersal, home range size, territoriality, mating system, and the nature of social interactions in a variety of species.

UNDERSTANDING OURSELVES

So much is similar in the basic biology of vertebrates, and

so universal are the processes of evolution, that an understanding of nonhuman sociality is likely to illuminate human society.

This point was hitherto neglected, but stressed by Tinbergen in the foreword to Kruuk's book, in which he concluded, "It is therefore imperative for the healthy development of human biology that studies of primates be supplemented by work on animal species that have evolved adaptations to the same way of life as ancestral man."

Following Wilson's Sociobiology, it has been widely and sometimes controversially discussed. Clearly, two routes come to mind as fruitful sources of this insight: looking at the societies of species most similar to our current condition (including some hypersocial aspects that put us in circumstances to which we have not yet had time to evolve) and focusing on those currently entering evolutionary phases through which we have already passed.

The first approach has prompted (or at least its promise has funded) much primatological research. The closeness of this parallel might be diluted if, as Hinde suggested, the societies of humans differ from those of other animals in that social structure in nonhumans is determined primarily by the sum of the interactions of its component individuals, whereas in human groups a structure is more often imposed from above by government or tradition in the form of Dawkins's memes.

Hinde's dichotomy may imply that the imposition of structure can cause stresses in human social systems when natural roles conflict with assigned roles. On the other hand, one could take the view that the dichotomy is not profound because the constraints of governmental ideology are loosely parallel to those imposed on all species by ecological factors such as resource dispersion.

If so, a different understanding of social responses to the imposition of external constraints might be revealed by species more recently launched onto a trajectory of sociability. Certainly, badger groups show weight reduction, higher incidences of wounding, and lower reproductive success per breeding individual as group size increases.

For badgers, group living may be a social innovation facilitated by the development of agriculture; individuals may be evolving towards capitalizing on this newly imposed structure (by manipulation, support, interdependence of roles, and other factors), but for the moment the stress is showing.

HOW TO DESCRIBE SOCIAL DYNAMICS

It does not detract from the excitement of behavioral, ecological, and sociobiological insights to note that recent enthusiasm for these topics has been characterized by a plethora of short, snappy papers with a clear adaptive punchline and a concomitant neglect of the empirical foundations of ethology.

Historically, this arises because behavioral ecology and sociobiology were pioneered to offer ultimate functional explanations, whereas ethology embraced adaptive significance and evolution along with mechanisms and ontogeny. This vogue has led to the widespread abandonment of the ethological aspirations of the late 1970s, epitomized by Hinde's careful use of terminology and hierarchical classification to ensure compatibility between studies used for comparative work.

At its purest, this traditional ethological approach placed greater emphasis on the facility of later reinterpretation of results than on the quest for a desired result. In contrast, there is an invasive tendency to treat hard data and description as disposable assets sacrificed to analytical elegance and discussion. In a science in its infancy (such as social biology) this brings the risk that future research may be doomed to repeat previous field work solely to attempt reinterpretation of undisputed results.

ACTION, INTERACTION, AND RELATIONSHIPS

Adapting Hinde's classification, the basic units of social exchange between individual primates are action and interaction. Actions are directed toward the environment, and are often important in understanding a previous or subsequent interaction with individuals. Interactions (A attacks B) are hard to interpret without records of actions (A accidentally drops fruit, B picks up fruit).

Above action and interaction in the hierarchy of social dynamics are social relationships. Relationships are quantified by the rates, frequencies, and patterning of component interactions and may be described in terms of the diversity of interactions, the degree of reciprocity or complementarity, relative frequency and pattern of interactions, synchronicity, and multidimensional qualities.

In principle, relationships may be stationary or transitionary. The former do not change with prior experience (intrinsic development) or conditions (extrinsic modification) and are at most rare and perhaps nonexistent in mammalian societies. Generalizations about relationships can be sought in various ways.

Dyads may be assigned to predefined categories such as age, sex, kinship, or even personality. Personality, in this context, is a consistent moderator of interactions; for example, a shy animal tends to act differently from a bold animal under the same circumstances. Block model methods allow subgroupings to be isolated from sociomatrices.

For example, Iacobucci compared 13 methods for recovering subgroup structure from dyadic interaction data. In general, block models are based on structural equivalence of sociomatrices. Some individuals behave similarly in respect to their age, status, or sex. Using block models, these relationships between individuals can be extracted and further studied, for example, using tests for reciprocity and interchange of behaviors or using more detailed methods based on time structure of the processes.

In their study of capybara mating systems, for example, Herrera and Macdonald disentangled the effects of dominance on mating success. In that example, dominant males secured more matings than any other individual, but fewer matings than subordinate males as a class; this arose because while the dominant was busy driving off one subordinate, another sought quickly to mate with the female.

The sum of social relationships may be compiled in a

matrix of dyadic interactions to produce a social network. Analysis of sociomatrices assumes stationarity, which, as we have noted, is effectively nonexistent. The solution is to divide sociomatrices into appropriately defined periods that approximate stationarity.

This might involve consideration of, for example, "the first 50 interactions," "the second 50 interactions," and so forth, or "wet season interaction" versus "dry season interaction," or "simultaneous presence of dominant" versus "absence of dominant." Nested analysis is a common way to achieve this goal.

SOCIAL STRUCTURE, FROM SURFACE TO DEEP

A social network provides a snapshot of one facet of society. By analogy, one analysis of a social network is akin to the view through one window into a large and labyrinthine house; the view through all windows gives the structure. It is therefore necessary to compile several networks that describe different facets of one society.

This task may be made harder because different sociometric variables (e.g., grooming, aggression, and play) may not follow similar patterns of stationarity; aggression may covary with age and presence of dominant, whereas grooming may not. Notwithstanding these complexities, a society's structure can be described in terms of these networks.

Indeed, there are layers of completeness to this description. The structure that prevails may vary on a circadian basis, or seasonally or annually; it may also be characteristic of a species' society in only one habitat or set of environmental conditions. At its most fundamental, elements of social structure may characterize all populations of a species.

Therefore, social structure might usefully be categorized on a continuum from surface structures to deep structures. The study of social dynamics seeks to describe and explain the patterning of transitions and stability of social structures. An important goal of evolutionary biology is to identify the rules, derived from a variety of empirical and theoretical sources, that are thought to guide an individual's decisions in a social context—Axelrod's seminal question of whether to cooperate. An accurate description of structure is clearly a prerequisite to a sensible exploration of these rules.

Operationally, the point at which a thorough description of social structure is complete is probably the first point at which it is legitimate to consider exchanging data language (grooms, fights) for theory language (alliance formation, competition). These interpretative substitutions are topics for the discussion section of a paper, whereas in the results section data should be presented without such interpretation.

Exploring exhaustively the layers of structure in animal societies is a major undertaking. Not least because a major objective of studying social dynamics is to contribute to the solution of practical conservation problems, it is inevitable that such studies may sometimes have to be undertaken quickly.

The obvious shortcut, within the framework of existing theory, is to use selected revealing behaviour as a guide to an overall system. As a loose analogy, this is akin to using the seating arrangement at a formal dinner as a guide to the social role of the guests in contexts beyond the dinner. Naturally, such shortcuts necessitate validation.

BEHAVIORAL PARAMETERS

Even to a casual observer it is obvious that most behaviour patterns occur in bouts; that is, they do not occur randomly in time. Although the existence of bouts may be obvious, how best to define them is not. A review of major textbooks and many papers reveals that many prefer to avoid this issue.

The most common usages define bouts as a repetitive occurrence of the same behavioral act (states or events) or a short sequence of behavioral actions that occur in some functional pattern. States usually have durations, and states with extremely short durations are called events.

The difficulties of defining the hierarchy of bouts, states, and events is illustrated by allogrooming by a mouse. This comprises a series of actions (nibbles) of short duration; an uninterrupted string of these nibbling actions might make up a bout of grooming.

However, a student of the detail of mouse grooming will see that these strings of nibbles may sometimes transfer from one body region to another (e.g., from head to neck to flanks or back), and for some purposes it may be helpful to distinguish bouts at this finer scale (a bout of head grooming distinct from one of flank grooming). The problem is that the best definition of a bout depends on the purpose of the analysis in which it will be used.

Depending on the scale of resolution required, even a short sequence of nibbles at one patch on the flank might be distinguished from another bout of grooming at the next patch of fur. Ultimately, each nibble could be defined as a state, punctuated by another state (shifting the head a fraction to grasp the next tuft of fur). At a given level of resolution it may be helpful to define states from which bouts are built up, but often, under closer scrutiny, a state will emerge to have a structure that could itself represent a bout (rather than one nibble, or one sweep of the paws, while grooming).

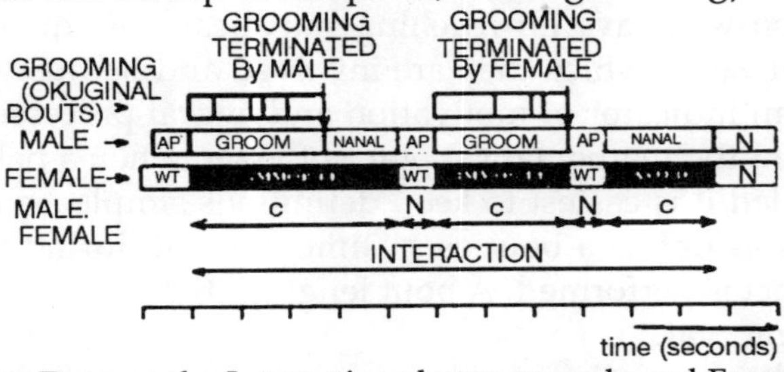

Fig. Data on the Interactions between male and Female Wood mice, Illustrating the Arbitrariness in Defining a Bout.

In practice, the best definition of a bout depends on the purpose of the analysis. The top row illustrates that grooming (of the female by the male) is split into a series of brief nibbles partitioned by fleeting changes of position. Each period of nibbling might be defined as an event or as a bout.

If each nibble is an event, then the sequence of them might make up a bout of grooming. Bouts might also be defined in terms of contact between the two mice, and in that case the period during which the male was first grooming and then nasoanal sniffing the females constitutes one bout of contact, during which the female was immobile. Finally, the entire period of male—female interaction might be defined as a bout.

This scheme is based on real data on the behaviour of wood mice. Depending on the model under analysis, grooming bouts could be distinguished as one continuous process or as a series of original grooming bouts in which a mouse shifts between body positions. Even an interaction can be considered a bout if several criteria are fulfilled. C = contact behaviour, N = noncontact behaviour, AP = approach, WT = wait, N-anal = nasoanal contact.

In some contexts this wracking down of the microscope to reveal more and more detail may seem merely a quest to bring into view the number of angels perched on the pinhead, but an important point nonetheless emerges: the unit of much behavioral analysis is the bout, and the usefulness of a definition of a bout is affected by the level of magnification at which the analysis is being undertaken.

Bouts must be defined very carefully because their definition will have far-reaching statistical consequences for any analysis in which they are involved and because of their role as an indicator of motivation and neural processes.

From the mathematical point of view, when a behaviour is modeled it is easiest to keep definitions simple, so Haccou and Meelis define a bout as a "time interval during which a certain act is performed. A bout length is the duration of such a time interval."

In the calculation of transition matrices, transitions to the same act are impossible, so diagonal elements in the matrix are treated as zero (the notion of transitions from one bout of behaviour to another of the same behaviour presupposes that the observer can distinguish a bout from an interval between two consecutive bouts).

The complexities of defining bouts seem particularly difficult to accommodate in studies of mammals, for which one action pattern often appears to slur into another. For example, although the task of defining a bout of drinking by a fox (made up of a string of lapping events) is feasible, the parallel quest of distinguishing the gulps of a rat or primate (which drink continuously) might be dubious. In comparison it would be straightforward, in a parallel study of a bird, to identify consecutive pecks and the interval between them, which can be analyzed using log—survivor plots.

The awkward truth is that the convenient hierarchy of bout, state, and event is often arbitrary and often concerns a continuum of lengths (clearly revealed by techniques such as the Noldus videotape analysis system, which measures even the durations of events).

To continue with the example of allogrooming wood mice and opting for Haccou and Meelis's definition of a bout as either the duration of a state or the interval between two states, it can be straightforward to recognize bouts. Wood mice, for example, often switch between two states during exploration: scanning while walking and rearing up or scanning while immobile.

In this example, the behavioral elements scan, rear up, and walk each have durations and are therefore states. For example, the duration of scanning is a bout. During scanning, however, wood mice often turn their heads back for only a tenth of a second; turn-head-back is an event. In this case, one bout may include scanning interrupted by several turn-head-back events.

Sometimes, however, it is difficult to distinguish between bouts, although methods are available to do so. When behaviour is studied in sequences (i.e., continuous records), the existence of bouts can be confirmed because the distribution of their lengths follows an exponential distribution as long as successive bouts comply with the assumptions of a first-order continuous-time Markov chain model. In practice, if bout durations do not follow an exponential distribution, there are two possible explanations.

First, the observer incorrectly recognized the bouts and therefore measured the wrong thing, perhaps because some bouts were only partially observed through insufficient time for observation. Second, the first-order Markovian assumption is not upheld. In the latter case, there are again two possibilities. First, bout lengths may exhibit dependency (between successive bouts or every second or third bout). In this case it is always better to use a semi-Markovian model. Second, and more abstruse, there may be second- or third-order Markov models, which according to Haccou and Meelis are rare and intractable in ethology.

In general, bouts approximate to a first-order Markov process and therefore can be distinguished on the basis of testing their distribution or fitting a nonlinear curve to the logarithm of observed frequencies of gap length. Because most behavioral studies are based on parameters such as latency, duration, and time intervals, bout length is a very important parameter to study at the outset.

STATIONARITY

Obviously, if circumstances change over a series of bouts it is confusing or misleading to lump them for analysis. Consequently, it is important to identify stationarity, during which things do not change. Plots of transition frequencies from one act to another in several consecutive periods can be used to judge whether the behavioral process is approximately stationary. Haccou and Meelis emphasize that nonstationarity can mask treatment effects and make the results of an analysis ambiguous.

Quite apart from the mathematical implications of inhomogeneity in the data, the existence of the motivational change that causes nonstationarity may be the very object of study, so it is important to identify it from an ethological standpoint. The complexities of stationarity reverberate through ethological methodology. Bekoff, for example, is skeptical that the concept of stationarity can be applied in a social context and therefore concludes that Markovian analysis should be avoided.

THE ETHOGRAM

Early in any study a researcher must classify the behaviour patterns to be documented. This classification is an ethogram, and constitutes a dictionary of the researcher's language; without an ethogram, meaning is not fixed.

Ethograms have become unpopular because (like dictionaries) they take up large amounts of space and are dull to read. However, they are vital reference material for those wishing to assess a study's conclusions critically.

As a corollary, it is not uncommon for the brunt of reviewers' comments now to fall on statistics rather than data acquisition, reflecting the same shift in emphasis away from data, and onto interpretation, as the hard currency of behavioral science. By analogy, consider the shift from real objects to paper representations in financial systems: 50 years on, would you rather find a stash of gold or war bonds?

We propose that the Internet, with its capacity to archive and disseminate large quantities of multimedia information, will revolutionize the use and usefulness of ethograms. For example, in parallel to a written publication on badger vocalizations, we have created a multimedia vocal ethogram and another based on digital video.

Ethograms are often constructed in a pretrial period during which behaviour appropriate to the study is chosen and described. However, short studies are unlikely to detect a complete catalog of relevant behaviour and are likely to miss social behaviour characteristic of context-sensitive dyadic interactions. Fagen tentatively proposed that a sample of 50,000 acts or more might generally be necessary to estimate the repertoire of a typical carnivore or primate species.

Fagen's estimate was based on the biologically general type—token relationship. This relationship was shown by May to be a general statistical property of all catalogs in which an infinite number of objects belong to a finite number of categories. It prescribes that the logarithm of the number of types in an inventory depends approximately linearly on the logarithm of the total number of acts in the catalog. In practice, our badger study confirmed the point:

Some antagonistic behaviors (coordinated attack on the rump of an animal after a partner has grasped its head) that we predicted might occur from their occasional appearance in ritualized play between cubs were seen for the first time in earnest between adults only after thousands of hours of direct observation. Although rare in badgers, these coordinated attacks gave an important insight into a previously undocumented realm of cooperative aggression.

What are the components of an ethogram? Ideally, each element might be purely descriptive, free of any imputed function. A heroic attempt at this purity of description was Golani's use of balletic choreographic scores to quantify the spatial and sequential organization of body part movement, together with its qualities such as speed and force and the degrees of variation tolerated within categories. It was intended to help solve what Golani called a blind spot in the behavioral sciences: the need for a universal language to describe animal movement.

Despite potential for universality of description, such quantitative choreography has proved impractical for most field studies. Also, focusing on the minutiae of postures may be too detailed for this purpose. A realistic option, facilitated by the Internet, is the creation of an archive of film clips and spectrographic catalogs within a taxonomic library.

However, such a "content and quality" ethogram is still denuded of context. Is a bird pecking a conspecific attempting fighting, grooming, or feeding? The answer may seem obvious from the rate of pecking, its strength, and the reaction of the other bird, but logging these contextual clues is time-consuming.

This is why many ethograms incorporate a shorthand notation that uses not only purely descriptive language but also proximate function language, derived from context and likely result. The demons of tautology and teleology lurk in such language.

BEWARE TELEOLOGY

Omnipresent dangers in the study of social behaviour are the closely allied traps of teleology and unwitting anthropomorphism. Teleology, or the doctrine of final causes, infers purpose in nature (e.g., to infer the existence of a creator from the works of creation). The teleological conviction that mind and will are the cause of all things in nature is not within the scope of scientific method. In the context of animal behaviour, teleological explanation would name and account for a behaviour by its presumed ultimate effect (e.g., appeasement, submission, or punishment) and not by its proximate causes or physical appearance.

It may be convenient to label as "punishment" the category of attack launched by a dominant meerkat on the only member of her group not to join in a fight with territorial trespassers. The danger lies in (inadvertently) interpreting the functional nuance of this convenient label as the proximate cause of the attacker's behaviour.

We know only that one individual did not join the fight and that the dominant member of its clan then initiated an attack on it (joined by all its group-mates). It is a matter of interpretation whether this attack functioned as punishment and a matter of speculation whether the attacking meerkat had punishment (or anything else) in mind when it attacked; it is certainly unwarranted to conclude that the putatively punitive attack was launched with the ultimate purpose of increasing the fitness of the attacker (although that may well be its consequence).

Interpretation based on context again harbors the pitfalls of premature use of proximate function language. It would be folly, for example, to categorize as "supportive" an instance of a large female grooming subordinate kin but to label as "repressive" the grooming by the same female of nonkin subordinates. In such a case, proximate function language would have prematurely slipped all the way to ultimate function language.

Clearly, the risk is that the very hypothesis used to make this interpretation may later be said to be supported by the observation; in this hypothetical case, the erroneously circular conclusion would be that kin selection theory is supported by the observation that females are supportive of kin and repressive to nonkin. In fact, such theories can be tested only by following actions through to fitness consequences—a very difficult task.

In the progress of a given scientific inquiry, there is no

virtue in premature exchange of data language for theory language. The teleological trap arises because in humans, conscious action (and even empathy to other people's unconscious actions) often involves mental plans and objectives that stimulate or explain the performance of behaviors. But consciousness has not been proved in other species and to confuse proximate factors with ultimate factors in this way may stifle correct interpretation.

As Kennedy noted, "Teleological terms such as 'searching' ... really describe the animal's presumed state of mind. ... Trying to find an objective substitute for a teleological term will always pay off in research because it forces a mental break-out and a closer look at the components of the behaviour actually observed, components which the handy teleological term leaves unnoticed or at best unformulated in the back of the mind."

Kennedy concluded that scanning would be a more useful term for searching. The worst cases of teleology may use such "catchy" labels that proximate and even ultimate function are obscured in the mind of the reader (if not the writer) because of the human innuendo of the word. Rape, used as a label for resisted matings, has been widely considered a case in point.

Used correctly, "mock anthropomorphism" can be a very valuable heuristic tool to guess the function of behaviour, using our own mental processes to explain behaviour in an ultimate fitness context. Such hypotheses can be tested, but even if they are supported we must beware of unwittingly crossing the threshold of assuming that the animal has used the same mental processes when deciding on its own actions.

The temptation of teleology—to impose our mental model on others—may be greatest when dealing with other primates. Tinbergen foresaw this when he wrote, "Teleology may be a stumbling block to causal analysis in its less obvious forms. ... The more complex the behaviour systems we deal with, the more dangerous this can be." These strictures against teleology do not imply that ethologists diminish animals to the level of Descartean machina anima; rather, the objective is to be mindful of Lloyd Morgan's canon that.

In no case may we interpret an action as the outcome of the exercise of a higher psychical faculty, if it can be interpreted as the outcome of the exercise of one which stands lower in the psychological scale". This is a methodological rather than ideological stance, which may be more likely to be correct than the converse assessment.

As Dawkins (1989:95) put it, a pragmatic aid in navigating this tricky terrain is to advance "always reassuring ourselves that we could translate our sloppy language into respectable terms if we wanted to."

CLASSIFICATIONS OF BEHAVIORAL INTERACTIONS

How, in practice, can we impartially assign positive or negative implications to behavioral interactions such as proximity or grooming that are ambivalent, without recourse to the very theories we seek to support? One solution lies in sequence analysis, using unequivocal behaviour patterns as anchors; high-intensity attacks are certainly detrimental to the receiver's fitness, whereas allowing mating is usually beneficial to fitness. If transition probabilities revealed grooming as a predictor of attack, then it might usefully be classed as aggressive, but if grooming often precedes acceptance of mating, it is probably amicable.

Firm grounding of a behaviour in context may justify an interpretative classification. Often, however, it is impossible to establish a uniform link between ambivalent behaviors (e.g., grooming) and anchors (e.g., coalitionary aid or attack) until the whole social network of interactions within the group has been described. Descriptions based on ultimate function should be eschewed until such links are established.

A very long list of behaviour patterns or relational types could be described and analyzed at each level of the structural hierarchy of social dynamics. Here, we consider just one example at each of three levels: grooming as an interaction, dominance as a relationship, and the concept of "social group" as a structure.

Grooming: Allogrooming can be abundant, observable, and

clearly bidirectional or unidirectional, a combination of attributes that have made it perhaps the most frequently studied social interaction. Despite the common assumption that it is amicable or beneficent, it may also cue either dominance or submission in a ritualized context.

The style of grooming, be it simultaneous, alternating, or routinely one-sided, is often largely species typical and even a small sample may give important clues to social relations in a wider context. Grooming has been the focus of many primatological studies, among which it appears to have more to do with social bonding than with hygiene.

For example, there is a significant correlation between time spent grooming and group size but not body size. The idea that primates compete for grooming access in a social setting has also been very influential.

To unravel more complex intricacies requires more probing analytical techniques, of which an especially rigorous example is Hemelrijk and Ek's study of reciprocity and interchange of grooming and agonistic "support" during conflict in captive chimpanzees.

Reciprocity was defined as one act being exchanged for the same thing (grooming for grooming), whereas interchange was defined as involving two different kinds of acts being bartered (i.e., grooming for support). They created an actor matrix (who initiates to whom) and a receiver matrix (who is recipient from whom) and used a correlation procedure, the K r test, to examine links between the two matrices while taking account of individual variation in tendency to direct acts.

They also attempted to retain stationarity in the data by distinguishing between periods when an alpha male was clearly established or when alpha status was in dispute. Hemelrijk had shown that grooming and support were both independently correlated with dominance rank, so whenever reciprocity or interchange was part of a triangle of significant correlations, they partialed out the third variable to see whether the remaining two retained significance.

Between the females, for example, a significant interchange between grooming and support received did not depend on dominance rank. Females groomed individuals that had supported them regardless of rank, indicating a social bond. However, the reverse was not true in that support was not given in return for grooming, so it was not possible to "buy" support with grooming. Patterns of relationships were different between the sexes.

Cheney used the distribution of grooming among individuals within a group to examine whether adversity, in the form of rivalry between groups, had a uniting effect within the group; the expectation of this hypothesis was a more egalitarian spread of grooming within the group when competition with neighbors was intense.

This expectation was not upheld, perhaps because neglected individuals would suffer a greater loss of fitness by shirking in their support of their own group than by attempting to trade this support for better treatment, even assuming a lack of punishment for failure to cooperate.

Even when intergroup competition was strong, marked hierarchies in grooming favoritism characterized intragroup relations. This study used the distribution of grooming to generate data with which to infer intragroup cohesion and equality. The ultimate functional explanations given depend directly on the validity of that assumption.

Our acceptance of such hypotheses may be influenced more by our intuitive anthropomorphic bias for their seeming correctness than by our purely objective assessment of the supporting evidence. This is the dilemma of being a social mammal attempting to describe the societies of other mammals.

The site to which grooming is directed may have social implications. Among long-tailed macaques (Macaca fascicularis), for example, low-ranking females expose the chest, face, and belly less often to higher-ranking females than vice versa. Male—male dyads almost never groomed the face, chest, and belly. Similarly the face, chest, and belly were underrepresented in grooming between individuals that groomed each other less frequently.

One interpretation is that recipients of grooming expose

less vulnerable parts of their body to, and avoid eye contact with, individuals they perceive as potentially dangerous.

Another analysis of the form, distribution, and context of social grooming was undertaken in two groups of Tonkean macaques (Macaca tonkeana) by Thierry et al. They found that kinship and dominance role had no effect on the form or distribution of social grooming among adult females, the most common class of individuals to be observed grooming.

The caveat that no one method of disentangling social relationships is universally appropriate has been stressed by Dunbar. For example, in a study of grooming in three species of macaques, Schino et al. found that whereas frequency and total duration of grooming were highly correlated, there was low correlation between its total duration and mean duration, and frequency and mean duration were not correlated.

A closer inspection of the data also revealed that even this general equivalence between measures of frequency and total duration did not always hold within specific grooming relationships. In general, frequency is a measure of initiation and mean duration is a measure of continuation.

In Schino et al.'s case, both measures were needed to interpret adequately the macaques' society. Not surprisingly, one important stimulus for grooming in rodents is soiled fur. However, because grooming can be undertaken alone or mutually, it has social implications transcending cleanliness.

Stopka and Macdonald (in press) used a similar approach to Hemelrijk and Ek's analysis of grooming in captive chimpanzees to examine patterns of reciprocity in grooming among wood mice (Apodemus sylvaticus). A powerful method for investigating this involved row-wise matrix correlation and the Mantel test. Allogrooming was less common than autogrooming and was most commonly directed by male wood mice to females.

Having described a social network in grooming, Markov chain analysis then revealed that the termination rate of female contact behaviour (expressed as her tendency to flee the male's attentions) depended largely on the male's tendency to terminate allogrooming. The pattern of transition rates from one behaviour to another revealed that the most common transition was from allogrooming of the female by the male to male nasoanal contact with the female. This indicates the motivation underlying the male's tendency to allogroom females: males groom a female in order to gain the opportunity to sniff her nasoanal region and, ultimately, to decrease her tendency to flee, thereby allowing them the opportunity to mate.

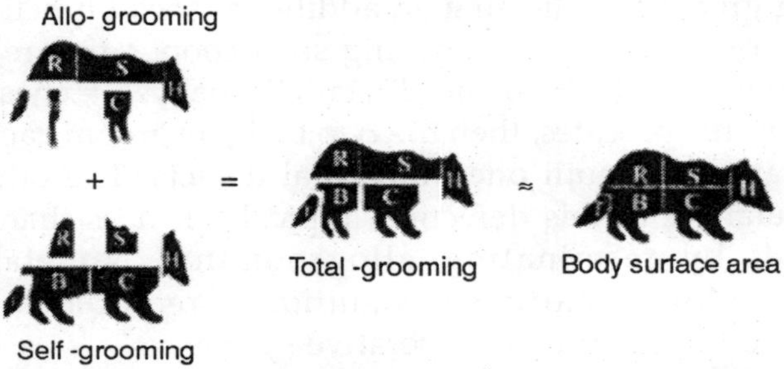

Fig. Silhouettes Represent the Proportion of Time, per Square Centimeter, Spent Grooming Different Portions of the Body Surface.

Females are most likely to mate with males that groom them the most. This constitutes an unusual example of the market effect in a short-lived mammal, with females essentially selling sex in exchange for grooming. Allogrooming is the clearest and most common form of cooperation among badgers.

It appears to be a valuable supplement to self-grooming because it is focused on the parts of a badger (back of neck, shoulders, and upper rump) that are most difficult for an individual to reach for itself. The proportion of grooming directed to each segment of the body differs, but is complementary between self-grooming and allogrooming.

Allogrooming cooperation is maintained in this case by the simple expedient of ensuring that as much grooming is given as received. Adult badgers groom one another simultaneously using a responsive rule set (Axelrod and Hamilton 1981) that dictates that grooming can be initiated generously, but rapidly withdrawn (at a mean of just 1.2 seconds) if the recipient does not respond in kind. It shows that although self-grooming or allogrooming alone is strongly biased toward particular body regions, a combination of self-grooming and allogrooming produces a fairly even body coverage. This supports a generally utilitarian function for badger grooming.

The slight overrepresentation of the shoulder is the result of that being the most common site of grooming initiation during greeting, indicating an additional social function or a necessary social etiquette during such cooperative relations. R = rump, S = side, H = head, T = tail, B = belly, C = chest. If the recipient reciprocates, then the two badgers groom each other simultaneously until one individual defects. The other will then retaliate to this defection, typically in less than half a second, by terminating allogrooming. No elaborate scorekeeping or partner recognition is required to guard against cheating in this cooperative system.

We hypothesize that the bartering of allogrooming is so direct because in the putatively primitive society of the badger, individuals spend most of their waking hours foraging separately from one another (Kruuk 1978). As a consequence, there are few opportunities to pay back beneficence in other currencies such as coalitionary aid. Perhaps more importantly, there are also few opportunities for badgers to exert manipulative pressure and extort grooming from other individuals by threat of negative reciprocity. An exception may help prove the rule:

During the breeding season, when larger badgers can despotically control breeding opportunities around the sett, smaller individuals sometimes abandoned the "You scratch my back, I'll scratch yours" strategy and groomed larger individuals even without reciprocation. The only other exception to the direct reciprocation rule was found when mothers groomed cubs that were too young to reciprocate grooming, a behaviour that doubtless accrues fitness benefits in terms of cub survival for the mother.

DOMINANCE

Terms such as dominance are used to describe predictable

aspects of repeated dyadic interactions (i.e., relationships). The term dominance is a valuable concept made controversial largely by a proliferation of definitions; Drews, having reviewed 13 definitions, concluded that dominance is characterized by a consistent outcome in favour of the same individual within a dyad, within which the opponent invariably yields rather than escalating the interaction.

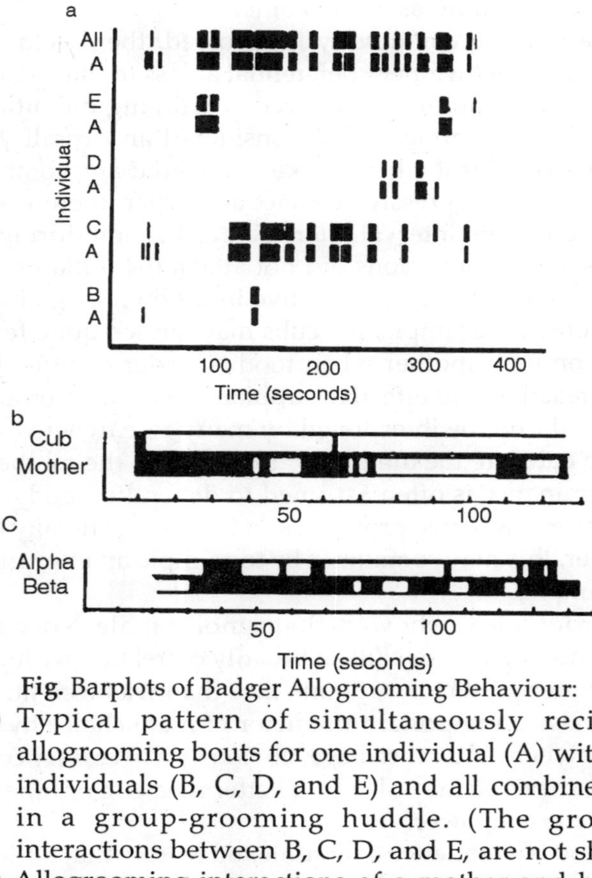

Fig. Barplots of Badger Allogrooming Behaviour:

(a) Typical pattern of simultaneously reciprocal allogrooming bouts for one individual (A) with other individuals (B, C, D, and E) and all combined (All) in a group-grooming huddle. (The grooming interactions between B, C, D, and E, are not shown.

(b) Allogrooming interactions of a mother and her cub, showing the characteristically oneway beneficence.

(c) Grooming of one adult male to a large adult female, one of the small minority of observations in which allogrooming between adults is not simultaneously reciprocal.

This excludes many uses of dominance and, in particular, does not allow the winner of a single agonistic interaction to be defined as dominant. A dominance hierarchy may be produced by ordering the dominance relationships between dyads. There may be a problem in determining what constitutes yielding as an outcome.

If a limiting commodity is involved, then yielding may be allowing the "winner" deferential access to the commodity. Where no commodity is involved, a working definition may be used. For example, Chase considered an overall yielding response to occur if a hen chicken delivered any combination of three strong aggressive contact acts when there was more than a 30-minute interval after the third action during which the receiver of the actions did not attack the initiator.

A certain degree of subjective intuition may go into such definitions; for example, fox cubs may launch quite ferocious attacks on their mother when food is under dispute, but the lack of reaction and effective supplanting of the mother at the food would not be interpreted by many researchers as a sign that the cubs are meaningfully dominant to the mother.

Dominance is often assumed to depend on body size, as a proxy for combative prowess, and this may generally be true. However, this may sometimes be too simple an interpretation. For example,

Berdoy et al. showed that among male Norway rats (Rattus norvegicus) age (itself broadly correlated with weight) was a better predictor of dominance than was weight. Age of male is also an important predictor of mate choice by female spotted hyenas. Advanced age, obviously a necessary corollary of good survival, may also be a better measure of fitness than current musclepower.

There is a tendency to assume too readily that all societies are arranged in a straightforward dominance hierarchy. Our own work with badgers revealed some of the complexities of elucidating and interpreting the dominance concept. We were interested in feeding interactions because the pattern of food availability and its manner of exploitation are believed to be central to badger social organization.

We investigated feeding dominance by establishing artificial food stations in the field. This experimental approach was deemed necessary because although dominance interactions may be key components of a social system, they may also be too rarely expressed to investigate with statistical rigor in a purely natural context. If a longer period is spent accumulating data, the requirement of stationarity may be breached and dominance relations may change during the observation period.

Artificial provisioning experiments allow wild social groups with settled relationships to serve as subjects while improving the number, quality, and rate of observations. There are clear perils to the approach, however; ethically we had to ensure that injurious levels of aggression were not provoked, and scientifically the general relevance of the experimental protocol had to be verified in an unmanipulated context. For this reason we maintained surveillance and control over the experiments using a live infrared video link and pursued further related observations in different social contexts.

We found that when badgers were presented with a single food source requiring contest competition for access, there was little direct evidence of default yielding to certain challengers and hence strict dominance relations: a feeding badger generally escalated aggression to some degree against any challenger. However, the study revealed predictors of contest success, so that we could predict relative supremacy in a dyad. These predictors of supremacy included length and weight and, to a lesser degree, age.

These attributes alone were not sufficient to predict the direction of supremacy in a dyad, however, because the role of challenged feeding individual also proved to be associated with contest success. And even when an individual occupied the same role against the same adversary, we still could not fully predict the contest winner; subsequent rechallenges often had a different outcome, indicating that some motivational factor such as hunger can also play a decisive role.

Hence, in the badger's case it did not prove helpful to attempt construction of a dominance hierarchy, linear or otherwise, from the provisioned patch data. A measure of relative supremacy based on an individual's characteristics and role in a contest proved to be more heuristically valuable.

To further complicate matters, it became clear that if more food sources were provisioned at one site, so that contest competition was no longer necessary for feeding access, scramble became the preferred method of competition. This cast doubt on the relevance of our predictors of relative supremacy derived from contest competition because most natural badger foods are presented in a way that does allow such scramble competition.

Interestingly, however, the behavioral patterns of contest we observed during feeding competition (e.g., side-to-side flank barging) were also seen in male—male competition for estrous females. Analysis of those data revealed that weight, length, and age as well as prior ownership were also predictors of copulation access, leading us to believe that these assets may indeed have general relevance in predicting the outcome of competitive dyadic interactions among similarly motivated badgers.

Social Groups ition may go into such definitions; for example, fox cubs may launch quite ferocious attacks on their mother when food is under dispute, but the lack of reaction and effective supplanting of the mother at the food would not be interpreted by many researchers as a sign that the cubs are meaningfully dominant to the mother.

In terms of their constituent individuals, a group may be transient, as in herds of wildebeest, or nearly permanent, as in packs of wolves. Some may even move daily between these extremes; in the dry season, stable groups of capybaras may coalesce at water holes into transient herds.

Many groups can be defined spatially, in terms of the proximity that members maintain to each other. Within territorial species, cohabiting occupants constitute a spatial group, and may be brought together by limited or variably available resources or by sociological advantage.

Membership of spatial groups can be assigned using spatial criteria, within which indices of association can be derived from dyadic proximity interactions; Macdonald et al. report systematic patterns in the frequency with which members of a farm cat colony sit adjacent to other individuals. Among these cats, and among the aforementioned capybaras, the social status of some low-status individuals is defined by their spatial peripheralization.

As Martin and Bateson noted, extra meaning can be given to such criteria if it can be shown that there is some aspect of synchrony or complementarity between the actions of individuals in each defined group or dyad relative to outgroup individuals. Clutton-Brock et al. used a behaviour similarity index to quantify such complementarity. At a broader scale there may even be physiological synchrony within groups or between clusters of groups, as within packs of dwarf mongooses or between packs of Ethiopian wolves.

Spatially coherent groups can include a variety of types. Brown's schema, adapted by Lehner, is useful in that context. It identified five broad categories of group: kin groups (families and extended families), mating groups (pairs, harems, leks, and spawning groups), colonial groups (e.g., nesting sea birds), survival groups (groups drawn to each other for reasons of foraging, herding, or huddling) and aggregation groups (formed by physical factors acting on animals to force them into one geographical location).

Although membership of a spatial group is a frugal expression of nonrandom associations, a social group should exhibit some persistent convergence of behaviour or unified purpose that results directly from the association and interaction of the constituent network of individuals. That is not to say there will be no competition, antagonism, or discord within the group.

It may be endemic. But to differentiate a social group (e.g., a pride of lions) from an aggregation (brown bears at a salmon run may be an example), there must be some element of cooperation (positive interaction) to offset the disharmony. Some populations of badgers are organized into groups that cohabit within the same territory, develop social relationships, and include close kin, yet for a long time there was scant evidence that they accrued any sociological benefit from their association.

However as more has been learned of the social system of badgers, more potential advantages of social living have been found. These include winter huddling to conserve energy, sporadic alloparental behaviour (Woodroffe 1993), and occasional examples of coalitionary aid.

We have even observed young individuals apparently using the main sett as an information centre and trailing older adults to new feeding grounds after weaning. The important point to emerge was that none of these advantages appeared sufficient to explain social grouping, particularly because at the same time disadvantages of group living have been identified.

Rather, these social advantages appear to be secondary benefits derived from, but not fully explaining, their group living habit. On one hand, this spatial congruity is distinct from the aggregation of jackals or bears at a carcass or salmon leap, and perhaps even distinct from the community of antagonistic monogamous pairs of maras (Dolichotis patagonum) at a communal breeding den.

On the other hand, the badger spatial group is equally distinct from the social group of African wild dogs, whose cooperative hunting or collective defence of prey leads sociologically to per capita advantage.

Furthermore, the jackals meeting at a carcass may come from many different territories, but they may nonetheless meet frequently and have well-established social relationships. Indeed, as Macdonald and Courtenay showed, neighboring territorial canids may even be bonded by familial ties. It is therefore difficult to formulate a precise definition, and social group risks becoming a nebulous concept.

A social group constitutes a network of variously close affiliations: closely connected subclusters of individuals comprising subgroups, and groups connected by primary links between core individuals defined as supergroups.

Clusters preferentially and mutually exchanging services, support, or aid may be said to display friendships, and where clusters act in concert against others we may define them as coalitions. Of course, the benefits an individual can accrue through associations are not always mutually beneficial, and Kraft et al. designed a diving-for-food experiment to illustrate theft in rat society.

Rats were trained to dive into water in order to obtain food from another compartment. All individuals learned to dive; however, those that were able to steal effectively from their diving companions often opted for this less arduous means of securing food.

Macdonald et al. distinguished a hierarchy of questions to be tackled in quantifying cat sociality, and these could be adapted to a simple description of any society. First, on the basis of their individual comings and goings, what are the probabilities of a given dyad of cats being available to each other for interaction?

Second, when they are simultaneously present, what proximity do they maintain? Third, what is the overall frequency of interactions between them, and how does that translate to a rate per unit time in association? Fourth, what are the qualities of interaction? Fifth, what are the directions of flow from initiator to recipient for each type of interaction?

First, indices of association raise the question of what level of proximity constitutes being together, and the answer may vary between species from physical to visual or olfactory contact (Martin and Bateson 1993). CluttonBrock et al. classified red deer within 50 m of each other as in the same party, whereas farm cats generally position themselves within 1 m of the nearest neighbor.

Indices of association can be calculated in which x is the number of observation periods during which A and B are observed together, ya the number of observation periods during which only A is observed, yb is the number of observation periods during which only B is observed, and yab is the number of observation periods during which both A and B are observed but they are not together.

In some studies, as Ginsberg and Young pointed out, the proportion of individual A's time spent with B [x/(x + ya + yab)] and the proportion of individual B's time spent with A [x/(x + yb + yab)] may differ only because of a viewing bias concerning ya or yb. However, in the case of the farm cats the value x + ya + yab differs from x + yb + yab because individuals were genuinely present at the resource centre at different frequencies.

This leads to a problem with the simplest association index [x/(x + ya + yb + yab)] because A's association with B is very strong, whereas B's association

Fig. What Constitutes Proximity Between Individuals Differs Between Species and May Take Account of Physical Contact, Sight, Sound, and Smell.

Farm cats within a matrilineal group may cluster closely, but with individuals positioned in different annuli with respect to the most central females, whereas two red deer stags might be interacting intimately over a distance much wider than that separating entire societies of cats.

Observations may be biased by the observer's success in seeing them or by genuine differences in individuals' attendance records, but the time for which the two cats are simultaneously present defines the opportunity for interacting in the calculation of the example association index.

with A is not necessarily so, and vice versa. For example, if cat A is at the resource centre 40 times out of 40 scan censuses and cat B is there 10 times out of the same 40 scan censuses, and they actually sit together during 8 scans, then according to the simple association index, their association index is [8/(8

+ 30 + 0 + 2)] = 8/40 = 0.2. However, looking at this another way, it is also true that while at the resource centre cat A sits with cat B only 20 percent of its time, whereas cat B spends 80 percent of its time sitting with cat A.

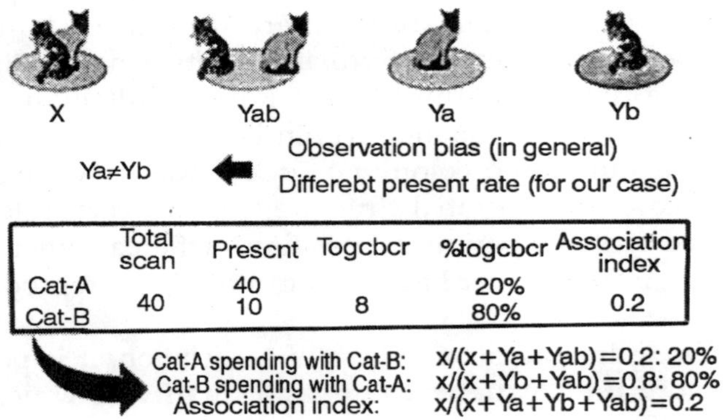

Fig. Considerations In Scoring Indices of Association are Illustrated by the Comings and Going of Farm Cats at a Resource Centre.

This asymmetry is accommodated if we calculate an association index for each cat rather than for each pair of cats. Thus cat A's association with cat B is $[x/(x + ya + yab)] = 0.2$, whereas cat B's association with cat A is $[x/(x + yb + yab)] = 0.8$.

Kerby and Macdonald and Macdonald et al. took scan samples of presence and proximity of cats at 30-min intervals at three colonies (small, medium, and large memberships), and made ad libitum records of the social interactions of focal cats. Data were summarized for each cat each month, and statistical analyses used these individual summary scores to avoid problems associated with pooling data.

Some generalizations spanned all colonies: the sexes did not differ significantly in their average monthly rate of presence, whereas age classes did; juvenile cats had the highest average monthly rate of presence, and adult males the lowest. Furthermore, each sex—age class displayed a favourite nearest-neighbor distance in terms of its tendency to maintain a given level of proximity to other cats.

At the large colony, when kittens were within 1 m of an adult female, this was their mother on 71.1 percent and 58.3 percent of occasions for male and female kittens, respectively, whereas adult males tended to be within 1 m of male juveniles most frequently and of male kittens least frequently. Despite such generalizations, there were also significant differences in the pattern of proximity between the colonies.

For example, at the large colony, adult males were significantly more frequently within 1 m of adult females, but at the medium-sized colony no such tendency emerged. A further corollary of spatial structure was that certain females occupied spatially peripheral positions, whereas others lived centrally, and the latter had significantly higher reproductive success.

Taking into account indices of association, it is possible to explore the frequency, rate, and quality at which cats interact. Take three cats, cat A spending twice as much time in the company of cat B as of cat C.

If A rubs on B and C 40 and 8 times respectively, and attacks them 10 and 2 times respectively, then A rubs on and attacks B at only 2.5 times the rate at which it rubs on or attacks C, although A rubs and attacks B five times as often as C. The quality of A's relationship with B and C is the same in that 80 percent of its interactions with both are rubbing and 20 percent are attacking.

In reality, at Kerby and Macdonald's large colony, individual central males averaged 602 (±254, SE) initiations relative to 502 (±142) for peripheral males, but the latter were present so seldom that their hourly rate of interaction when present averaged 6.9 (±1.5), in comparison to the central males at 2.3 (±0.7).

The situation was opposite for females: Peripheral females interacted at low frequency, and when their presence was taken into account, they interacted at an even lower rate. Furthermore, in the context of grooming,

Macdonald et al. showed that whereas the flow of licking differed significantly between pairs of interacting cats, it was invariably symmetric. In marked contrast, the flow of rubbing, though also differing between dyads, correlated with aggression but not grooming and was generally highly asymmetric.

They concluded that in the society of female cats, rubbing as a measure of status flowed centripetally toward central dominant individuals; grooming, which was exchanged reciprocally, was a measure of social (and often genetic) alliances, but unrelated to status.

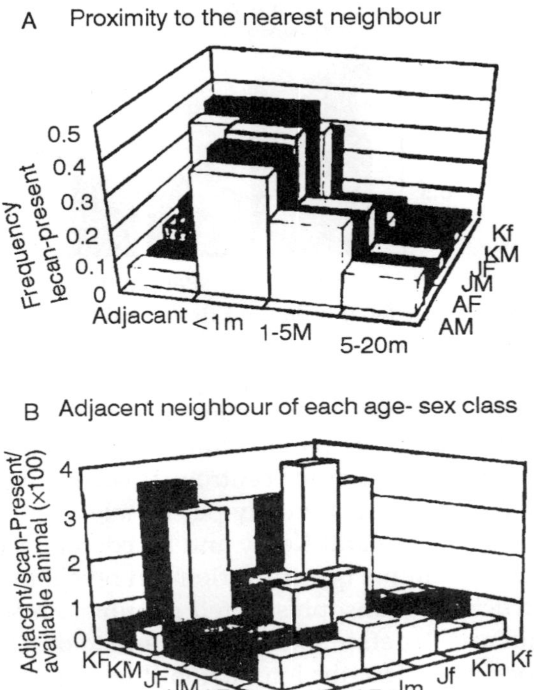

Fig. Exploration of Patterns of Spatial Proximity can Reveal some Unexpected Structure to what Appears Superficially to be a Random Gathering.

In the case of farm cats, these graphs, based on G. Kerby's field study in Macdonald et al. (in press), reveal (A) that different age—sex classes of individuals positioned themselves at significantly different distances to their nearest neighbors and (B) that these positionings differed significantly depending on the age—sex class to which that neighbor belonged. AF = adult female, AM = adult male, JF = juvenile female, JM = juvenile male, KF = female kitten, KM = male kitten.

-360-

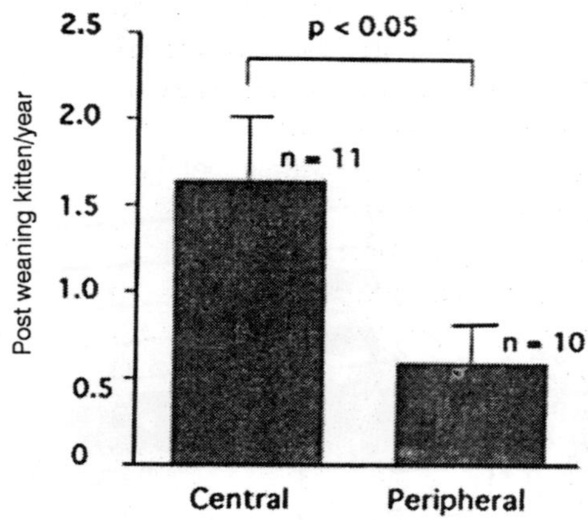

Fig. An Ultimate Goal of Many Ethological Descriptions is to Translate Indices of Social Behaviour Into Evolutionary Consequences.

In the case of Kerby and Macdonald's and Macdonald et al's farm cat study, the spatial arrangement of female cats around the farmyard resource centre was correlated with the number of kittens raised annually per female.

Macdonald et al.'s and Kerby and Macdonald's cat study shows how even simple quantification can reveal unexpected layers of structure in unsophisticated mammalian societies.

In summary, to define social groups and to describe social dynamics one must describe interactions that may be positive, tolerant, or negative in terms of their consequences for those involved. However, the choice of interaction type, or of the way of quantifying it, may radically affect the researcher's interpretation of the outcome.

This should not surprise us. An analogy with human social dynamics shows us that very different patterns of interaction appear if we view exchange of such commodities as money, conversation, or affection. And the picture changes yet again if we look at frequency rather than quantity or quality of the exchange. The most pernicious problem lies in correct interpretation of the context of the interchange. Is the individual to which the most money is observed to be given beloved kin, despised extortionist, or scarcely known shopkeeper?

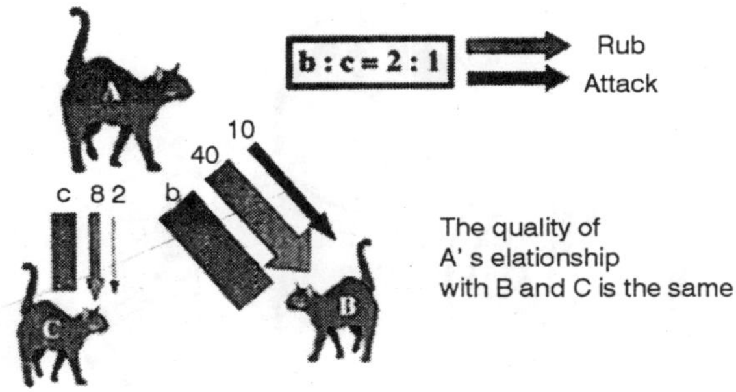

Fig. The Same Observations of Social Interactions Can be Expressed in Three Ways: As The Total Flux of A Given Behaviour Pattern, As A Rate, Or As A Proportion.

In this case, the total frequency is greater between A and B than between A and C (b > c), but qualitatively the components of their relationship are the same.

METHODS FOR BEHAVIORAL MEASUREMENT

Although our goal is not to provide an encyclopedic guide to the formidable practical problems faced by the field biologist, we briefly review several methodologies relevant to measuring social dynamics.

IDENTIFYING THE INDIVIDUAL

From the variety of ways of identifying individuals reviewed by Stonehouse, it is generally preferable to use natural characteristics. This necessitates validation of identification skills and it is imperative that behaviour is not inadvertently used as a cue to identity. For scientific as well as ethical reasons, it is also important to be alert to any influences of a marking technique on subsequent behaviour.

For example, in an analysis of a new fur-clipping technique used to mark badgers for individual identification,

Stewart and Macdonald assessed the effect of the clips on body condition using a matched-samples test. Our concern was that the technique could cause thermoregulatory disadvantage, and no significant effect of clipping on condition was found.

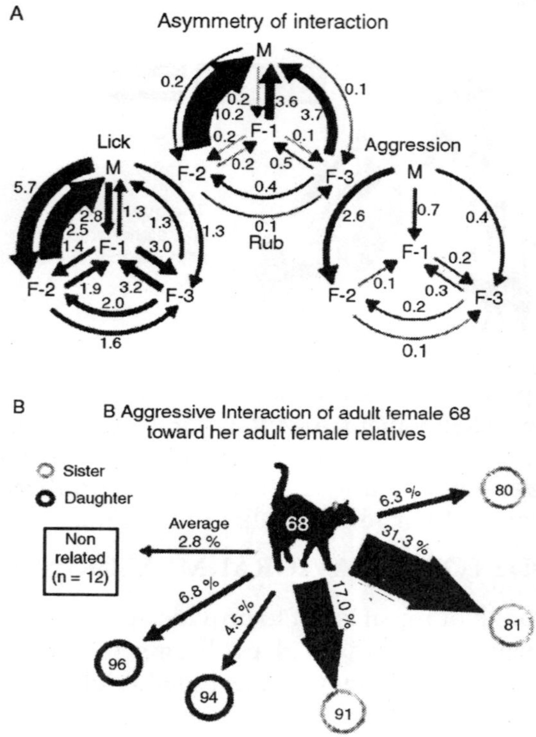

Fig. The Flow of Rubbing between a group of four Cats

Furthermore, marked individuals were capable of attaining high status in their group, as measured among males by copulation frequency during female estrus. It is logically impossible to be certain that clip marks had no effect on behaviour, but as in the case of Hoogland's Nyanzol dye-marked ground squirrels, we could detect no evidence that it did. However, absence of evidence is not evidence of absence, so caution is nxecessary. For example, bird rings or ear tags have been demonstrated to affect judgments of asymmetry and mate choice in some species.

This topic is further discussed by Bekoff and Jamieson. In a similar way, names of animals should be carefully chosen to avoid biasing behavioral observation of social role. The name given to an animal, be it that of a colleague, friend, or personality, can inadvertently influence all but the most objective observer. Even ascribing "male" and "female" names can cause recording biases.

Numbers are better (although "A1" may still find deferential recording to "C5"), but can be rather hard to remember. In our own badger study we primarily used names derived from the appearance of the identifying fur clip mark, giving asexual and fairly connotation-free names such as Line, Dot, Dash, and Corners.

SAMPLING AND RECORDING RULES

Sampling rules dictate which subjects to watch and when. Martin and Bateson distinguish ad libitum, focal, scan, and behaviour sampling. Recording rules specify how the behaviour is to be recorded. The two options are continuous recording and time sampling. Time sampling may be undertaken on an instantaneous or one—zero basis.

The qualities of each sampling rule are easily envisaged through the eyes of contrasting onlookers at a football match. The typical spectator engages in ad libitum sampling and samples everything. He or she is likely to record the goals and any spectacular fouls, with a bias toward the fouls committed by certain star players and those committed on the ball. In contrast, a talent scout might be a focal sampler, observing one player at a time.

He or she may gain a poor impression of the game as a whole, but a clear impression of the qualities of particular players. Of course, the scout must be alert to biases such as the stage of the match at which a given player becomes the focal subject (it would be inappropriate to compare one player observed in the first 5 min with another observed in the final 5 min).

A team manager might opt for scan sampling, censusing the behaviour of each player in turn, at rapid intervals, to get an overall impression of how the team functions as a unit. Finally, a gambler, interested only in the outcome of the game, may undertake behaviour sampling, recording only goals. By focal or scan sampling one could miss these events.

Turning to recording rules, a continuous record contains the times at which each behaviour started and ended, and thereby reveals true frequencies and durations. Instantaneous sampling characterizes the current behaviour at predefined intervals. One—zero sampling records whether a behaviour happened in one particular interval of defined length.

The most common combinations are focal sampling with a continuous record, scan sampling with an instantaneous record, and focal sampling with a one—zero record.

AD LIBITUM SAMPLING

An obvious bias in ad libitum sampling would arise if a researcher's presumptions led to particular individuals being disproportionately represented in the sample, thereby distorting their importance. This effect has been credited with the overestimation of male roles in the societies of some primates and may subsequently have been countered by contrasting biases.

Furthermore, individuals or behaviour patterns that are less conspicuous (because of short duration, subtlety, or association with obscured locations) tend to be underrepresented by ad libitum sampling. For example, Bernstein demonstrated that using ad libitum sampling of rhesus monkeys (Macaca mulatta), the frequencies of contact aggression and avoidance were underestimated whereas that of chasing was overestimated.

A related and insidious bias may affect the initial choice of study group. Sharman and Dunbar noted that studies of baboon behaviour tend to involve larger than average groups. This erodes the generalizability of the results because the behaviour of individuals was systematically influenced by the size of the group.

In the same vein, perhaps the bias of primatologists' expectations for their own taxon has lead to primates being described as having peculiarly complex social lives. However, other taxa also emerge to have intricate societies, an obvious example being the Carnivora. Despite this family of biases, ad libitum sampling can be useful during the naturalistic observation phase of pilot studies and is valuable as an adjunct to more rigorous protocols that might otherwise exclude important events that fell beyond their scope.

FOCAL SAMPLING

Because the whole attention of one observer is likely to be absorbed by one focal subject, focal sampling risks losing valuable contextual and simultaneously comparable information. The temptation to follow one focal individual for longer than is necessary reduces the effective data set.

Having observed a focal chimp for 300 hours, Kawanaka discovered that subsets of 25 hours reliably and consistently represented the same time budget. Excessive recording of particular focal individuals also worsens the task of finding periods of relative stationarity. Cycling through a series of focal individuals and then checking an individual against itself in an earlier period may circumvent this.

Video recording can sometimes relax these limitations. There may be a general warning in the finding of Arnoldmeeks and McGlone that focal animal sampling of one individual among a litter of young pigs provided a good measure of the time budgets of other piglets in the same period for all but social behaviour.

TIME SAMPLING

ONE—ZERO SAMPLING

One—zero sampling consistently overestimates duration and gives unreliable information on frequency. This can be partly overcome by reducing the recording interval to periods shorter than the bout length of the behavioral categories, but the technique then often becomes more cumbersome than continuous focal sampling. Therefore, this once common sampling method has largely been abandoned.

Using a cumulative Poisson process, it is possible to assess, for a given interval length, the probability that the necessary conditions will not be met for accurate frequency counts and unbiased duration estimates in one—zero sampling.

Suen and Ary also identified a post hoc correction procedure that produces results with negligible systematic errors in one—zero duration estimates. This correction procedure requires that more than five 0 scores lie between two consecutive but not adjacent 1 scores.

Despite its general fall from favour, one—zero sampling has some potential virtues. It gives a measure of the probability that an observer will see at least one example of the response in the specified time interval. Bernstein argues that this probability may be relevant to an animal's perception:

For example, when an individual is deciding whether to spend the next 5 min in the vicinity of another individual, the probability of attack may be more important than the mean duration or frequency of attacks. Martin and Bateson also note that the onset, duration, and frequency of some behaviour, such as play, are hard to represent with continuous sampling and hard to score during true instantaneous sampling.

The judgment that play has at least occurred in a previously defined interval is more likely to be accurate. Finally, some behaviour, such as copulation, may occur often in one short interval and then not be seen for many subsequent intervals. One—zero scan sampling reveals how long a researcher is likely to have to watch the animals to see an instance of such behaviour.

INSTANTANEOUS SCAN SAMPLING

This technique scores what some or all individuals are doing at predetermined time intervals. It produces a reliable estimate of the total durations of activities through a dimensionless measure of the proportion of occasions during which the behaviour is encountered. Instantaneous scans may give poor information on frequency, although this shortcoming is minimized if the sample interval is small enough to catch even the shortest intervening period between recurrence.

However, as in the case of short one—zero measures, such short bouts become more inconvenient than focal sampling. Furthermore, the need for statistical independence of the data points requires that sample intervals not be too short.

For example, if 10 successive scans all fell within a single long bout of a particular behaviour, then they constitute a sample of 1, not 10. Such dependence would require that the data be transformed to a proportion on an interval scale for comparison rather than being used as discrete categorical scores (and therefore could not be analyzed in an interaction sociomatrix).

As a rule, Bernstein suggests that if the probability of scoring the same continuous state twice in succession is less than 0.05, then the two scans can be treated as independent. In practice this means that the interval between scans should exceed the mean duration of a state plus 1.96 standard deviations. Even long intervals between scans can provide a basis for accurate estimates of mean time budgets, but the increase in variance caused by longer sampling intervals may swamp analyses of covariation and produce type II errors.

The pros and cons of each approach, and the diverse aims of investigations, mean not only that no particular method is universally superior, but that even for one study an eclectic approach to sampling is likely to optimize data acquisition.

A mix of all occurrence sampling, scan sampling, and focal individual instantaneous sampling is possible. The results of all three are likely to be enhanced if decisions on the durations of intervals and focal sessions are judged from preliminary analysis of a continuous focal sample (e.g., made by using a portable video camera in the field).

TECHNIQUES FOR BEHAVIORAL MEASUREMENT

Simultaneous and continuous focal sampling of all individuals in a group might be the most accurate and informative combination of sampling and recording, but circumstances may render this combination unattainable. Technology is at hand to help. Computer-based event-recording software provides an automated string of actor, behaviour (time), and modifiers (such as recipient of interaction and quality of interaction). Currently, memorizing the keyboard is the way to avoid taking one's eyes off the subjects, but voice recognition software may soon allow input of data spoken in the correct syntax.

Various software packages already assist behavioral recording, and the most sophisticated, such as Noldus Observer, offer a variety of tools for manipulation and basic statistical analysis of the data and an easy Windows interface. Noldus also performs sequential and nested analysis and allows simultaneous input from an alternative digitized source (e.g., heartbeat or external temperature).

Two caveats concern widespread adoption of sophisticated software. First, it may encourage a lazy uniformity in data acquisition that stifles innovation in sampling and recording rules or analyses. Second, a researcher's new-found ability to record a plethora of behaviour and contexts may exceed his or her ability to frame hypotheses and manipulate and analyse the data.

There are circumstances under which video recording is an economical alternative to direct observation and offers some unique advantages. The video record can be played and replayed, allowing multiple passes over complex sequences at a resolution of up to 50 frames per second. With a microphone, complex vocalizations can be spectrographically analyzed in context. The video and sound sequences may be digitized and stored using a PC and video capture board; this allows efficient indexed archiving of raw data and use in electronically distributed ethograms.

Video can also be time-coded and used with event recording packages to facilitate accurate data entry or even allow automatic data entry for movement, interaction, and some simple postural categories. Video can be used to calculate interobserver and intraobserver reliability, including the drift in ethogram categorization which often occurs in long studies. Limited use of video in pilot studies can allow the disparity between different behaviour sampling techniques and a true continuous record to be assessed.

In automatic field use, video allows easy habituation for many species (including facility to use infrared light to provide illumination beyond the visual range of most mammals). Continuous surveillance of multiple focal sites becomes a possibility for even a single researcher, with the equipment operating in conditions that would be too arduous for the researcher to endure for protracted periods.

Against these undoubted benefits video equipment can still be initially costly, with problems of equipment reliability, accessibility, and security. Video also has a more limited field of vision than a human observer, and because it requires transcription after the event, it may increase the duration of studies. Stewart et al. describe the use and construction of automatic video surveillance units, and Wratten provides a general background to video techniques.

Finally, although the naturalist's preference may be to observe social behaviour directly or with video, this may be uneconomical or unfeasible. Examples of indirect measures include the frequencies of bite wounds, chewing of ear tags, and hunting patterns from carcass retrieval. Radiotracking has revolutionized the study of elusive mammals. Although initially used principally to plot movements, the use of radiotracking to supplement classic fieldcraft and as an aid to observation has long been advocated.

It is also possible to infer features of social behaviour from the dynamic interaction between the movements of radiotracked individuals. The practicalities of radiotracking are reviewed in Kenward and analytical considerations in White and Garrot. An increasing number of software packages for analyzing tracking data are available, such as Wildtrak and Ranges V.

ANALYSIS OF OBSERVATIONAL DATA

Context must always have been important in human interpretation of the behaviour of other animals. Even our own aphorisms (blood's thicker than water, do as you would be done by, and so on) reveal an intuition that presages modern conceptual insights regarding the forces fashioning animal societies, including our own. Today, biologists interpret behaviour within a context fortified by theories of cognition, games, trait-group selection, kin selection, reciprocity, and byproduct mutualism.

The traditional approach to observing animals, with roots embedded in the Aristotelian idea of ascribing properties to animals, could yield detailed ethograms that demonstrated the defining behavioral uniqueness of each species. Interspecific comparisons of these behavioral lexicons shed light on phylogenies even before the days of phylogenetic regression.

The quest for more robust tools with which to elucidate the shared properties between species has, in the last 20 years, contributed to the blurring of boundaries between studies of ecology, ethology, and sociobiology. Today, the basic questions posed by those concerned with the contextual interpretation of animal behaviour still echo Niko Tinbergen's four factors: causation, development, survival value, and evolution.

The classic sequence starts with observation to document the behavioral repertoire in context, and this forms a basis for experimental study of causation, development, and survival value and an attempt to reconstruct the evolution of particular units of behaviour. The approach rests on parsimony and reductionism.

Of course, there are different forms of reductionism, and the biologist's variant, as described by Konrad Lorenz in The Foundations of Ethology, may deviate from the physicist's method of general reduction.

Here, we consider some limitations of the rationales behind four concepts widely used in the study of social dynamics: statistical rationality, matrix facilities, lag sequential and nested analysis, and concept of uncertainty measures (Markov chain analysis).

STATISTICAL RATIONALITY

A society is the product of social flux among its members and between them and outsiders. Therefore, for many purposes the description of a society requires quantification of the interactive components of behaviour between members of, for example, a reproductive unit, group, colony, or population; these may be continuously or instantaneously recorded.

In the analysis of observational data, insufficient attention to methodology can result in a rift between statistical significance and biological relevance, whereas the correspondence between these two can be fostered by attention to sampling tactics and statistical sensitivity. An inadequately sensitive statistical method or model may fail to reveal the properties of an interactive behaviour, even if an appropriate sampling method was used.

The familiarity of two types of error has rendered neither rare. Type I error involves mistaken rejection of a null hypothesis (i.e., a false positive); type II error is failure to reject a false null hypothesis (i.e., ignorance).

As a general statement, scientific validity depends partly on the sensitivity of the tests involved (decreasing the probability of type I errors) and on the number of hypotheses tested (the accumulating building blocks of the argument adding to its objectivity and decreasing the probability of type II errors).

However, there is a tradeoff between these kinds of error; decreasing type I error can increase the risk of type II error. There is also a risk that even when the hypothesis has been tightly phrased, a test may pick up "departures" from H 0 that result from "noise" rather than real differences. Matloff describes how confidence interval analysis may partly solve these problems. An advanced account of these topics is given in Thompson and Krishnaiah and Rao.

MATRIX FACILITIES: ANALYZING SEQUENTIAL DATA

A matrix is any rectangular array of numbers (e.g., frequencies, number of transitions). If the array has r rows and s columns it is called an r by s (r × s) matrix. Thus for example,

$$A(3\times3)\begin{vmatrix} a11 & a12 & a13 \\ a21 & a22 & a23 \\ a31 & a32 & a33 \end{vmatrix} \text{ and B } (2\times3)\begin{vmatrix} a11 & a12 & a13 \\ a21 & a22 & a23 \end{vmatrix}$$

Where each row can represent the actor entry and columns

the receiver; that is, the value a11 represents frequency (or other value) of relations between the row descriptor (r1) and the column descriptor (s1).

Our task here is not to explain the mathematical principles of matrices, but to describe their uses in behavioral sciences. Interaction frequencies among individuals, presented as a sociometric matrix, are often the basis for analysis of social dynamics. Individuals are classed as actors or initiators in the rows (r) of the matrix and as receivers in the columns (s).

Other sociometric matrices include distance and association matrices; in these, each cell contains a symmetric (dis)similarity measure for the pair of animals indicated by the row and the column of the cell. The cells of a transition matrix contain the frequencies with which the behaviour indicated by the row (the preceding behaviour) is followed by the behaviour indicated by the column (the succeeding behaviour).

Relationships are often presented in terms of the dyadic interactions between two individuals of the group. Frequencies of interactions within the dyad during defined time intervals are then written in sociometric matrices. These types of sociometric matrices are often used to identify the dominance hierarchy among individuals based on such key behaviour patterns as initiating aggression.

One individual is defined as a dominant (the winner), the other as a subordinate (the loser). The strength of the relationship is expressed in terms of linearity. Han de Vries developed an improved test of linearity in dominance hierarchies containing unknown or tied relationships. This test is based on Landau's linearity index, but takes the unknown and tied relationships into account. An unknown relationship commonly arises because two members of a dyad simply were not seen to interact.

This can occur for at least four reasons: sampling may be inappropriate, linearity of the dominance hierarchy may be an artifact of experimental circumstances (e.g., dispersal cannot occur in captivity), the formation of coalitions may lead to mutual protection or mutual avoidance, and linearity may not be a feature of the group's social structure (ultimately because the ecological circumstances offer no selective advantage to linear dominance with regard to, for example, resource holding potential). However, papers acknowledging the first two categories of problem are inevitably few.

The problems of elucidating a hierarchy and revealing its consequences are illustrated by wood mice (Apodemus sylvaticus). Although some aspects of this species' behaviour can be studied in the wild Macdonald 1994), their small body size and large ranges thwart many studies and make them strong candidates for observation in arenas or seminatural enclosures. Bovet found that male wood mice of different statuses tended to be active at different times.

Therefore, they rarely met, which thwarts study of their encounters. Such problems can be partly circumvented by the use of sociometric matrices. Because it illustrates the use of matrices, we will go step by step through Stopka and Macdonald's (in press) sociometric study of wood mice. Based on videotaped interactions of nine mice observed over 3 months, the frequencies of transition between fight or chase and avoidance for all individuals in the colony were ranked.

Each of these matrices can be analyzed for linearity in the relationships between the mice. In the case of avoidance, this provides a statistic (Kendall's linearity index $K = 0.775$), which indicates that the null hypothesis that avoidance relationships are randomly distributed can be rejected in favour of the alternative that they are linearly ordered ($\chi^2 = 43.76$, $df = 20.16$, $p = 0.005$). Similarly for allogrooming ($K = 0.592$), there is significant linearity ($\chi^2 = 34.96$, $df = 20.16$, $p = 0.03$).

However, linearity in avoidance relationships is stronger ($p = 0.005$) than that in allogrooming relationships ($p = 0.03$), raising the possibility that linearity in grooming merely reflects that in avoidance.

The statistical tool to test this hypothesis is the Kr row-wise matrix correlation test. When applied to the avoidance and allogrooming matrices (in which the rows and columns are identically ordered), this reveals a strong negative correlation (p-right = 0.941).

Table. Matrix of Frequencies of Transitions from Fight to Avoidance (top) and Frequencies of Allogrooming Behaviour (Bottom) Among Nine Wood Mice

Receiver (avoid)

	M1	M2	M3	F1	F2	M4	F3	F4	M5
M1	*	52	180	9	18	192	11	66	26
M2	8	*	22	0	6	15	3	11	0
M3	10	15	*	6	10	18	4	10	16
F1	0	0	0	*	1	0	0	1	2
F2	1	1	6	0	*	6	2	8	7
M4	2	4	4	1	3	*	4	4	4
F3	0	2	3	0	1	0	*	0	3
F4	5	3	5	0	4	3	0	*	1
M5	0	0	1	0	0	0	0	0	*

Receiver

	M1	M2	M3	M4	M5	F1	F2	F3	F4
M1	*	1	0	3	0	62	15	124	13
M2	0	*	0	0	0	6	3	8	6
M3	0	0	*		0	19	4	2	14
M4	1	0	0	*	0	4	1	8	4
M5	0	0	0	0	*	0	0	0	0
F1	6	1	2	2	0	*	0	1	3
F2	0	0	3	0	0	0	*	0	0
F3	14	0	1	1	0	0	0	*	1
F4	1	1	0	0	0	1	0	0	*

Therefore, the dominance (avoidance) and grooming matrices exist independently. Tendencies to allogroom and avoid differ between males and females; in practice, lower-ranking males are rarely groomed and they avoid dominant males and are avoided by females, whereas females are the main recipients of grooming, which is mainly initiated to them by dominant males. This pattern led us to conclude that in order to obtain matings, male wood mice have to groom females.

LAG SEQUENTIAL AND NESTED ANALYSIS

Sequential analysis facilitates study of the temporal structure of sequences of events. A common variant is lag sequential analysis, which involves calculating frequencies of transitions between pairs of events within a certain lag in a time series.

The first event of each pair is called the criterion event (also called antecedent, X-event, or given event) and the second event is the target event (also called the consequent or Y-event). Lag sequential analysis allows you to answer questions such as "How many times is the (criterion) event Head Up followed by the (target) event Stare?"

The time window in which the transition from criterion to target event occurs is either a state lag or a time lag. In a statelag sequential analysis, a transition is counted from a criterion event to the first target event following this criterion event.

In time-lag sequential analysis, a transition is counted from a criterion event to the target event occurring within a specified time window following the criterion event.

All other events that are not defined as either criterion or target events are ignored in the analysis. Lag-sequential analysis differs profoundly from other approaches (such as log-linear models), which compare estimated expected frequencies with observed frequencies because these do not have a sequential property.

The only way to analyse changes of behaviour through time is in terms of transitions between the antecedent and the consequent behaviour. Lag sequential analysis is currently the only method that applies a contextual approach: Instead of analyzing the frequencies of different behaviour over a certain period of time, lag sequential analysis uses the frequencies of behavioral transitions.

However, the results of lag sequential analysis are rarely expressed in terms of the absolute frequencies of transition from one event to the others; instead, they are often written in probabilistic terms as transition rates.

The transition rates, written in transition matrices, can be calculated from the frequencies of transitions between preceding and consequent behaviour using equation 1.6 in Haccou and Meelis, which is formally written as a maximum likelihood estimator (MLE)—equation when duration of an act is constant:

$$\alpha_{As} = \frac{N_{AS}}{N_A} \times \frac{1}{X_A}$$

or

$$\alpha_{As} = \frac{N_{AS}}{N_B}$$

Where β A,B is the estimator of transition rates, N A,B is the number of transitions from the state A to state B, N A is the number of states A, and $\bar{x}_A$ is the mean bout length of category A. MLE values are written in the state-space representation (also known as flow diagrams), that is, a set of states and transition rates between them.

In corroborative analyses at least two methods are currently used to describe behavioral processes: multiple-matrix analysis and nested analysis. Multiple-matrix analysis involves the construction of several matrices, one each for a variety of measurable parameters of social flux (e.g., dominance hierarchy, mutual grooming, approach and avoidance, acts of support, copulations, and paternity).

The pattern of flow between individuals for each behaviour pattern can then be compared between matrices using, for example, row-wise matrix correlation (association) or linearity indices (de Vries 1993). This approach provides answers to questions such as "Are dominant males groomed more frequently than subordinate males, or do females given protection by other females reciprocate this support?" This can involve construction of several matrices and row-wise correlations between them.

Nested analysis, on the other hand, involves defining questions within matrices. Subsets (nested categories) of data are selected within the overall matrix. For example, a researcher might ask, "Is there a tendency for juvenile males to be more successful in contests with rivals when positioned close to their mothers?" To answer this, the researcher would analyse the outcomes of fights when key individuals were adjacent to, or distant from, their mothers.

Both multiple matrix and lag sequential or nested analyses may involve the manipulation of very large data sets, and software is available to make this task more manageable. PC programs such as the Observer event recorder lag sequential analysis and the Matman programme are designated for recording and analyzing such data.

SEARCHING FOR A BEHAVIORAL PATTERN (MARKOV CHAIN)

It is useful when analyzing sequential data to characterize the chance of one behaviour following another. If exclusive acts (events or states) in a sequence are independent of each other, then the distribution of bout lengths will follow an exponential distribution; such a sequence is said to have Markovian properties. Such a process is completely characterized by a set of transition rates, which are the probabilities per unit of time of switching from one state to another.

By definition, acts within such a sequence are independent of each other and of time. The termination rate of an act is constant (i.e., independent of the past) only if the duration of the bouts follows an exponential distribution. This fact allows the inference of social interaction because dependencies between the terminations of different individuals would imply that they exchange signals.

Indeed, most sequences are not Markovian because the probability that most behaviors will occur depends on earlier events in the sequence and other confounding variables. Generally, for a given behaviour, the probability of occurrence of any future behaviour in the sequence is altered by additional knowledge concerning the past.

The programme Uncert analyzes serial dependencies in sequential events using the method of Markov chains and expressed in terms of the uncertainty that one event will follow another.

For example, Hailman and Dzelzkalnz used Markov chain analyses to reveal that tail-wagging by mallard duck (Anas platyrhynchos) punctuated a long sequence of behaviour patterns, indicating that a display is about to start (loosely analogous to the role of a capital letter beginning a sentence).

PREDICTABILITY OF BEHAVIOUR

Measures of uncertainty can be a valuable tool in assessing the predictability of responses of individuals to others. Again, we explore an example from our work on wood mice (Stopka and Macdonald, unpublished data).

For example, all combinations of five male and four female wood mice kept under video surveillance in an enclosure with abundant nest boxes sometimes denned communally, although there was no thermoregulatory need to huddle. It was therefore surprising that cohabitants were often agonistic to each other.

One of the most common categories of interaction was approach and nose-to-nose contact, categorized by Bovet (1972a, 1972b) as amicable, but lag sequential analysis revealed that although this (nose-to-nose) behaviour pattern was indeed sometimes a precursor to flank-to-flank or nasoanal contact, on other occasions it was followed by attacking or chasing.

Therefore, nose-to-nose encounters are defined by their context within a sequence. Sequential analysis revealed that nose-to-nose contact was sometimes the prelude to an amicable encounter and sometimes to an aggressive one.

The crucial question is how to recognize mixed categories. Haccou and Meelis recommend: performing a global inspection of bar plots to ensure stationarity, testing the bout lengths of all acts in the ethogram for exponentiality (e.g., Darling's test), using cumulative bout length plots and the likelihood ratio change point test to reveal deviations from exponentiality.

Attributable to lack of stationarity, and devising a rational basis (in the case of nose-to-nose in wood mice, lag sequential analysis) on which to split problematic categories into new robust categories. Other methods of splitting the behaviour according to the bouts are based on fitting a nonlinear curve to the logarithm of the observed frequencies of gap lengths and on the use of likelihood ratio tests in helping to determine whether the data occur in bouts.

However, it is important that the statistical procedure does not obscure biological insight, so the new, split categories must make sense. For example, it is clearly sensible to split the instances of follow-B (individual B) that were preceded by fight from those preceded by copulation or intromission because both new categories of follow (A, B) have different contextual functions and, it turns out, different bout lengths.

Splitting of the category nose-to-nose contact did not make sense because it was equally likely to precede a fight or nasoanal contact. However, on further exploration, the probability that noseto-nose contact would lead to one or other of these categories depended heavily on the sex of interactant. Such sex-dependent sequences may even resolve puzzles in video surveillance by allowing inference of the sex of an unmarked individual.

SEQUENCES THROUGH THE MIST

All animals are social, even if only to the extent of engaging in sexual reproduction and otherwise avoiding each other. Understanding the rules that govern social dynamics in general is a step to understanding our own lives. Everything happens in context, so it is unsurprising that we advocate a contextual approach to studying society.

Social relationships are the consequences of entwining sets of events and states, and our aim has been to explore how these may be disentangled through analyzing sequences of changes in time and space.

Although some of these analytical approaches may seem daunting through their unfamiliarity, they concern a phenomenon— sequentiality—that not only is at the root of life in the patterning of nucleic acids, but also suffuses every aspect of our daily experience in strings of words and sentences and stories. One story can be told in different ways for different purposes, just as an individual butterfly is expressed differently as larva, pupa, and imago.

A metamorphosis in storytelling shows how the

perceptions of Robert Henryson, a fifteenth-century Scottish poet, might, for the purposes of measuring the dynamics of mammalian society, be stripped to their essentials. In his Moral Fables, Henryson penned these words:

This country mouse lay flattened on the ground, fearing every minute that she would be killed; for her heart was pounding with strokes of fear, and feverishly she trembled hand and foot. And when her sister found her in such straits, for the sake of pity first she grieved a bit, then comforted her with words as sweet as honey.

It is common for stories to be peppered with anthropomorphism, and one class of these, akin to mock anthropomorphism, enriches prose with insight. Certainly, understanding is not the sole prerogative of those who wield the colorless pen of late-twentieth-century science. On the other hand, another class of anthropomorphism is blind to life's processes and, in its fancifulness, more often than not corrupts literature by diminishing understanding.

Stripping out these distractions is a step toward understanding, and interpretation of the essence that remains hinges on the meaning of sequences, whether they be words or behaviors. In the contextual study of animal behaviour the narration that precedes or succeeds a given action has equal weight to the action itself. As a step toward such contextual analysis, we might wonder whether Henryson had this in mind:

One rural female wood mouse (Apodemus sylvaticus) adopted a dorsoventrally flattened posture, exhibiting, in an unspecified but seemingly fearful context, tachycardia and elevated body temperature, and associated tremulous movements of the fore- and hind-paws. These signs appeared to be stress-induced.

A conspecific female approached and initiated prolonged nonreciprocal grooming, perhaps with the consequence of diminishing the signs of putative stress. Subsequent molecular analyses confirmed the hypothesis that these individuals were siblings.

The only merit of the aridity of this version is that it

exposes the states and events that can usefully be defined in an ethogram and exposes the sequences of those that will become grist for analysis:

- *Ethogram*: crouch: dorsoventral flattened posture with head close to the (i.e., modifier) ground; shake: tremulous movement of body (no modifier), or shaking fore-paw(s) (modifier 1) or shaking hind-paw (modifier 2); approach: one individual moves toward another; groom opposite: one individual grooms the other on the head, or neck, or back, or flank (modifier 1).
- Sequence: Behaviour: A-crouch → A-shake → B-approach-A → B-nose-to-nose-A → B-groom-A.

Modifier 1: ground paw
Modifier 2: foot
Time: t 0 → *t* 1 → *t* 2 → *t* 3 → *t* 4 → ... *t* n;;
Duration: $(t_1 - t_0)$ $(t_2 - t_1)$ $(t_3 - t_2)$ $(t_4 - t_3)$ $(t_n - t_4)$
Subjects: 2 female wood mice (Apodemus sylvaticus)

Clearly, there are many ways of describing an observation. Henryson's original and our coded version may be quests for different sorts of understanding.

The point is that having designed an ethogram and having chosen the correct method of observation, we can distill an observed set of events or states into a pattern that facilitates greater precision of interpretation and decreased risk of fanciful anthropomorphism.

Chapter 9

Modeling Species Distribution

From the variety of checklists, atlases, and field guides available around the world it is easy to understand that distribution ranges are pieces of information that are seldom absent in a comprehensive description of species. Their uses range from a better understanding of the species biology, to simple inventory assessment of a geographic region, to the definition of specific management actions.

In the latter case, knowledge of the area in which a species occurs is fundamental for the implementation of adequate conservation strategies. Conservation is concerned mostly with fragmentation or reduction of the distribution as an indication of population viability, given that, for any species, range dimension is considered to be correlated to population size.

Unfortunately, animals move and this poses problems in mapping their occurrence. Traditional methods used to store information on species distributions are generally poor. Distributions have been described by drawing polygons on a map (the "blotch") to represent, with varying approximations, a species' range.

The accuracy of the polygons relies on the empirical knowledge of specialists and encloses the area in which the species is considered likely to occur, although the probability level associated with this "likelihood" is seldom specified. A more sophisticated approach divides the study area into subunits (e.g., administrative units, equal-size mesh grid), with each subunit associated with information on the presence or absence of the species.

In this case the distribution range of a species is defined

by the total of all subunits in which presence is confirmed; however, blank areas are ambiguous as to whether the species is absent or no records were available.

New approaches tend to overcome the concept of distribution range and move toward one of area of occupancy. This concept is particularly useful for conservation action and has therefore been included in the new IUCNRed List criteria.

For example, imagine a biologist who needs to find zebras. Intuitively, the odds of finding zebras in Scandinavia are very low, but moving to Kenya greatly increases the odds. This process is based on very basic assumptions such as that zebras live in warm places, say, with an average annual temperature of 13–28°C.

Obviously our observer won't expect to find zebras in every place on Earth that has an average annual temperature of 13–28°C; there are many other ecological requirements, along with other reasons, such as historical constraints and species behavioral patterns, that contribute to define the distribution of the zebra.

Nevertheless, if our biologist extends the same process, taking into account the preferred ranges of values of various environmental variables, the probability of finding the species in the areas in which these preferences are simultaneously satisfied increases.

If the aim of our researcher is to map the areas in which the species is most likely to be found rather than to find an individual, the entire process can be seen as a way of describing the species' presence in terms of correlated environmental variables.

And if inexpensive and broadly acquired environmental data (e.g., vegetation index maps derived from satellite data) are used to define species probability of presence, then maps of species distribution can be produced quickly and efficiently.

To provide a formal approach to species distribution modeling, the process can be divided into two phases. The first phase assesses the species' preferred ranges of values for the environmental variables taken into account, and the second identifies all locations in which these preferred ranges of values are fulfilled. The first phase is generally called habitat suitability index (HSI) analysis, habitat evaluation procedures (HEP), or, more generally, species—environment relationship analysis.

The second, which involves the true distribution model, has seen its potential greatly enhanced in the last 10 years by the increasing use of geographic information systems (GIS), which can extrapolate the results of the first phase to large portions of territory.

The power of GIS resides in its ability to handle large amounts of spatial data, making analysis of spatial relationships possible. This increases the number of variables that can be considered in an analysis and the spatial extent to which the analysis can be carried out.

Thus GIS provides a means for addressing the multidimensional nature of the species—environment relationship and the need to integrate large portions of land (eventually the entire biosphere) into the analysis to produce robust conservation oriented models.

Models and methods used in GIS-based species distribution models; it is based on a literature review carried out on GEOBASE with the following keywords: GIS, remote sensing (RS), wildlife, habitat, and distribution.

The 82 papers collected were classified according to the main tool used (GIS or RS), the modeling approach, the analysis technique, the discussion of the assumptions, and the presence of a validation section. At the same time, information was gathered on the use of the term habitat, the number of variables used for modeling, and the kind of output produced.

Far from being comprehensive, the review was the starting point for a tentative classification of GIS distribution models; at the same time, it allowed us to focus attention on some issues that we consider among the most important for correct use of GIS in species distribution modeling. In fact, although it offers powerful tools for spatial analysis, GIS has been largely misused and still lacks a clear framework to enable users to exploit its potential fully.

These issues range from unspecified objectives in the

process of model building to the lack of adequate support for the assumptions underlying the models themselves.

Before discussing these issues, we address the problem of terminology inconsistencies, which has a much broader extent in ecology than the specific realm of species distribution modeling. The problem emerges from our review and is probably caused, in this context, by misleading use of the same term in the different disciplines that have come to coexist under the wide umbrella of GIS.

TERMINOLOGY

Multidisciplinary fields of science are very appealing because they bring together people with different experience and backgrounds whose constructive exchange of ideas may generate new solutions.

In fact, many solutions that have been successfully developed and used in one field of science may, with minor changes, be used in other fields. The very nature of GIS makes it essential that specialists in different scientific disciplines contribute to the general effort of setting up and maintaining common data sets.

One drawback is that in the early phases of tool development (such as GIS), people who master the new tool tend to become generalists, invading other fields of science without having the necessary specific background. This may cause problems both in the solutions provided, which generally tend to be too simplistic, and in terminology, because the same term or concept can be used with slightly different meanings in different disciplines.

This is the case, for instance, with use of the concept of scale. For the cartographer, large scale pertains to the domain of detailed studies covering small portions of the earth's surface, whereas for the ecologist large scale means an approach that covers regional or even wider areas.

Obviously this derives from the fact that cartographers use scale to mean the ratio between a unit measure on the map and the corresponding measure on the earth's surface, whereas the ecologist uses it in the sense of proportion or extent. For example, the relationship between the geographic scale and the extension of ecological studies supplied by Estes and Mooneyhan highlights that large scale in ecology is often associated with small geographic scale:

Site = 1:10,000 or larger
Local = 1:10,000 to 1:50,000
National or regional = 1:50,000 to 1:250,000
Continental = 1:250,000 to 1:1,000,000
Global = 1:1,000,000 or smaller

In ecology it would be better to use the adjectives fine or broad, which places the term scale more in the context of its second meaning.

If the confusion arising from the two uses of large scale seems trivial (at least from the ecologists' point of view), we believe that the different uses that have been made of the word habitat give rise to major misunderstandings and thus need to be clarified.

HABITAT DEFINITIONS AND USE

The term habitatforms a core concept in wildlife management and the distribution of plant and animal species. The fact that the actual sense in which it is used is rarely specified suggests that its meaning is taken for granted. However,

Merriam-Webster's dictionary provides two different definitions and Morrison et al. observed that use of the word habitat remains far from unambiguous. The latter distinguished two different meanings: one concept that relates to units of land homogeneous with respect to environmental conditions and a second concept according to which habitat is a property of species.

Our literature review provided us with a variety of definitions and uses of the term habitat that are wider than the dichotomy suggested by Morrison et al.

We arranged these various meanings according to two criteria: whether the term relates to biota (either species and or communities) or to land, and whether it relates to Cartesian (e.g., location, such as a position defined by a northing and easting) or environmental space (e.g., the environmental envelope defined by factors such as precipitation, temperature, and land cover).

Although the classification in table allows us to partition the different definitions of habitat we have traced, in reality this partition is rather hazy. For instance, definitions range from the place where a species lives, which is a totally Cartesian space—related concept, to the environment in which it lives. In this last case habitat is seen as a portion of the environmental space.

At both extremes of the range of definitions, the slight differences in the terms used allows us to define a continuous trend between the Cartesian and the environmental concept, which is further supported considering a few definitions that combine the Cartesian and the environmental space. These last authors define habitat as the area that has specific environmental conditions that allow the survival of a species. Note that all of these definitions relate habitat to a species and some describe it as a property of an organism.

With a similar range of definitions, another group relates habitat to both species and communities. For instance, Zonneveld in accordance with a Cartesian concept, defined it as "the concrete living place of an organism or community." Others relate it to both Cartesian and environmental space, defining it as the place in which an organism or a community lives, including the surrounding environmental conditions (.

All of the definitions cited so far defined habitat in terms of biota. Zonneveld remarked that the term habitat may be used only when specifying a species (or community). Yet habitat has been used as an attribute of land.

Riparian habitat, for instance, is a specific environment, with no relation to biota. Use of habitat in this sense is widespread in the ecological literature. The concept predominates in ecology applied to land management such as habitat mapping, habitat evaluation, and habitat suitability modeling. A similar meaning of habitat is used in a review of habitat-based methods for biological impact assessment.

Although it has been used very often in this sense, we

were unable to find a single definition. A closely related concept, the habitat type, which is used in habitat mapping, has been defined as "an area, delineated by a biologist, that has consistent abiotic and biotic attributes such as dominant or subdominant vegetation".

Daubenmire noted that this meaning of habitat type corresponds to the land unit concept. In articles dealing with habitat evaluation, the term is used in a similar sense.

The use of an ambiguous term leads to confusion in communication between scientists. The ambiguity of habitat is also observed within the same publication. Lehmkuhl and Raphael, for instance, simultaneously used "old-forest habitat" and "owl habitat."

Even ecological textbooks are not free from ambiguity. Begon et al. (1990:853) defined habitat as "the place where a micro-organism, plant or animal species lives," suggesting that they consider habitat a property of a species.

However, when outlining the difference between niche and habitat, they later described habitat in terms of a land unit: "a woodland habitat for example may provide niches for warblers, oak trees, spiders and myriad of other species."

Confusion arises with respect to habitat evaluation as well. When defined as a property of a species, unsuitable habitat does not exist because habitat is habitable by definition. In this case some land may be classified as habitat and all of this is suitable. When defined as a land property, all land is habitat, whether suitable or unsuitable, for a specific species.

Why is the term habitat used in these various senses? The word originates from habitare, to inhabit. According to Merriam-Webster the term was originally used in old natural histories as the initial word in the Latin descriptions of species of fauna and flora. The description generally included the environment in which the species lives.

This leads to the conclusion that habitat was originally considered a species-specific property. It is interesting to note that the definitions we traced originated both from ecology and geography, suggesting that the confusion was not the result of separate developments in two fields of science.

At some time habitat started to be used as a land-related concept, most likely in conjunction with habitat mapping. A possible explanation for the change is given by Kerr (1986), who remarked that mapping habitat individually for each species would be an impossible job.

He argued that a map displaying habitat types and describing the occurrence of species in each type would be more useful to the land manager. This suggests that the land-related habitat concept arose because it was considered more convenient to map habitat types rather than the habitat of individual species.

We suggest that there was a second reason for the popularity of habitat type maps. In general the distribution of species is affected by more than one environmental factor. Until a decade ago it was virtually impossible to display more than one environmental factor on a single map. The habitat type, defined as a mappable unit of land "homogeneous" with respect to vegetation and environmental factors, circumvented this problem and was the basis of the land system (land concept) maps developed in the 1980s.

However, it is based on the assumption that environmental factors show an interdependent change throughout the landscape and that the environmental factors are constant within the "homogeneous" area. Thus to a certain extent the land unit meaning of the term habitat arose as a way to overcome operational difficulties in species distribution mapping.

Nevertheless, given that the variation of one environmental factor affecting the distribution of a species often tends to be independent of the other environmental factors, homogeneity is seldom the case, so there is seldom a true relationship between species and habitat types.

The advent of GIS has made it possible to store the variation of environmental factors independently and subsequently integrate these independent environmental surfaces into a map displaying the suitability of land as a habitat for a specific species.

The first examples of such GIS-based habitat mapping

were published in the second half of the 1980s Since then there has been a steady increase of the number of GIS-based habitat models. The increase illustrates a move away from the general habitat-type mapping applicable for multiple species toward more realistic species-specific habitat maps.

At the same time, the habitat type loses its usefulness because of the decreasing need to classify land in homogeneous categories. In other words, species-specific habitat mapping is increasingly incorporating independent environmental databases processed using information on the preferences of the species concerned.

In view of the anticipated move toward species-specific habitat models, we prefer to use the original species-related concept of habitat instead of a land-related concept; to avoid confusion, environment relationships and ecological requirements instead of the terms species habitat and habitat requirements.

GENERAL STRUCTURE OF GIS-BASED MODELS

The rationale behind the GIS approach to species distribution modeling is straightforward: the database contains a large number of data sets (layers), each of which describes the distribution of a given measurable and mappable environmental variable.

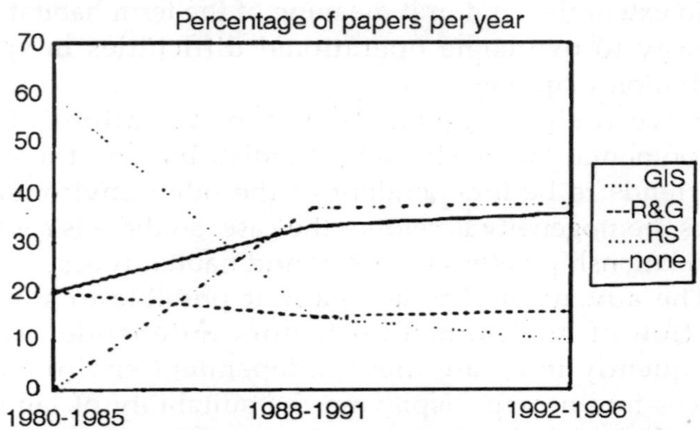

Fig. Percentage of the Papers Dealing with Habitat Modeling Using no Spatial Information.

The ecological requirements of the species are defined according to the available layers. The combination of these layers and the subsequent identification of the areas that meet the species' requirements identify the species' distribution range, either actual (if there is evidence of presence) or potential (if the species has never been observed in that area).

This basic scheme can be implemented using different approaches. A few classifications based on different criteria have been attempted. For example, Stoms et al. classified models based on the conceptual method used to define the species—environment relationship, whereas Norton and Possingham based their classification on the result of the model and its applicability for conservation.

Accordingly, Stoms et al. classified GIS species distribution models into two main groups—deductive and inductive—whereas Norton and Possingham gave a more extensive categorization of modeling approaches.

We have tried to define logical frameworks that can be used to classify species distribution models based on the major steps that must be followed to build them. To this end, we find the deductive—inductive categorization the most suitable starting point because it focuses attention on the definition of the species—environment relationship, which is the key point for the implementation of distribution models.

The deductive approach uses known species' ecological requirements to extrapolate suitable areas from the environmental variable layers available in the GIS database. In fact, analysis of the species—environment relationship is relegated to the synthesizing capabilities and wide experience of one or more specialists who decide, to the best of their knowledge, which environmental conditions are the most favorable for the existence of the species.

Once the preferences are identified, generally some sort of logical or arithmetic map overlay operation is used to merge the different GIS environmental layers to yield the combined effect of all environmental variables.

When the species—environment relationships are not known a priori, the inductive approach is used to derive the ecological requirements of the species from locations in which the species occurs. A species' ecological signature can be derived from the characterization of these locations.

Then, with a process that is very similar to the one used in deductive modeling but is generally more objectively driven by the type of analysis used to derive the signature, it is used to extrapolate the distribution model.

The data flow of GIS-based species distribution models for both the deductive and the inductive approaches. Whereas in the deductive approach GIS data layers enter the analysis only to create the distribution model, in the inductive approach they are used both to extrapolate the species—environment relationship and the distribution model. Along with the data flow, the steps that need validation are also evidenced in the figure.

Both inductive and deductive models can be further classified according to the kind of analysis performed to derive the species—environment relationship. Essentially these can be subdivided into two main categories: the descriptive and the analytical. Models pertaining to the first category use either the specialists' a priori knowledge (deductive—descriptive) or the simple overlay of known location of the species with the associated environmental variable layers (inductive—descriptive) to define the species—environment relationship.

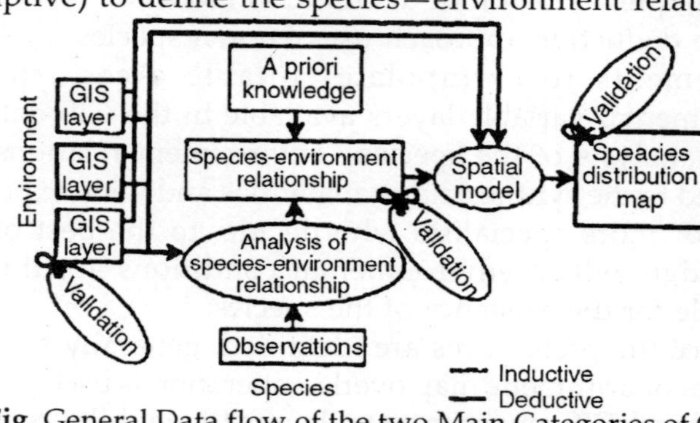

Fig. General Data flow of the two Main Categories of Gis

Descriptive models generally are based on very few environmental variable layers, most often just a single layer. They tend to describe presence and absence in a deterministic way; each value or class of the environmental variable is associated with presence or with absence (e.g., the species is known to live in savanna with an annual mean temperature of 15–20°C, so savanna polygons falling within the adequate temperature range are to be included as suitable environments).

No attempt is made to define confidence intervals to the individual estimate, nor is any information provided on the relative importance of one variable over another (e.g., vegetation types vs. temperature). Moreover, no estimate of the degree of association or its variability is provided with the relationship.

On the other hand, models that fall into the analytical group introduce variability in the sense that advice from different specialists is combined to define species—environment relationships, thus introducing variability in terms of different opinions of the experts (deductive—analytical), or that the species observation data are analyzed in a way that takes into account the range of acceptability of all environmental variables measured, their confidence limits, and their correlation.

Both the deductive—analytical and the inductive—analytical approaches tend to estimate the relative importance of the different environmental layers considered in the analysis, thus moving toward an objective combination of environmental variable layers.

Examples of deductive—analytical models are based on techniques such as multi-criteria decision-making (MCDM), Delphi, and nominal group technique (NGT). Generally speaking, these techniques use the advice of more than one specialist as independent estimates of the "true" species—environment relationship and evaluate its variability based on these estimates.

Inductive—analytical techniques rely on samples of locations that are analyzed with some sort of statistical procedure. Different techniques have been used, including generalized linear, Bayes theorem approach, classification trees, and multivariate statistical methods such as discriminant, discriminant barycentric analysis, principal component analysis (PCA), cluster analysis, and Mahalanobis distance.

Models that use simple univariate statistics, such as anova, Pearson rank correlation, and Bonferroni, pertain to a different subgroup because these analyses do not generally allow for definition of the relative importance of the environmental variables.

Further differences should be outlined for models that rely on the interpolation of density or census estimates to extrapolate distribution patterns. Although we have included these models in the inductive—analytical group, the geostatistical approach on which they are generally based suggests putting them into a slightly different subgroup.

Generally, discrete models are built associating the presence of a species to polygons of land unit types (e.g., vegetation categories), most often with a deductive approach; in fact, transferring into the realm of GIS, the traditional way of producing distribution maps is based on a similar but more arbitrary partitioning of the study area.

There are also some examples of binary classifications of continuous environmental variables (e.g., slope, aspect, elevation) using statistical techniques such as logistic regression or discriminant analysis. Categorical—discrete models do not account for species mobility and tend to give a static description of species distribution. Nevertheless, this approach can be used to address the problem of defining areas of occupancy and thus can be used successfully for problems of land management and administration.

On the other hand, probabilistic models can describe part of the stochasticity typical of locating an individual of a species and can be used to address problems of corridor design and metapopulation modeling, introducing the geographic dimension in the analysis of species viability.

We have considered GIS and RS as two different views of the same tool, the former being more devoted to spatial correlation analysis and the later more concerned with basic data production. In fact, the two families of software tools share many basic functions and are evolving toward integration into a single system. It should be noted that the review includes not only papers that use GIS or RS but also some that deal with HSI, HEP and general assessment of species' ecological requirements.

The papers in this last group do not generally represent examples of spatial models, in the sense that their products are not distribution maps, but they have been included because they are considered to be just a few steps away from a real distribution model. In fact, they describe the ecological requirements of the species in terms of mappable environmental conditions.

Most of the papers that use the deductive approach consider the a priori knowledge sufficient to define the ecological requirements of the species under investigation. This is especially true of papers that model distribution on the basis of interpretation of remotely sensed data; in fact, 15 out of 16 papers pertaining to the deductive group that used remotely sensed data to model species distributions fall within the descriptive group.

In these papers, image classification techniques tend to receive more emphasis, whereas the ecological application is most often seen as an excuse to apply a specific classification algorithm.

The time trend of the papers published shows rather stable use of RS technology and increasing use of GIS. Up to 1986, no paper makes explicit reference to the term GIS, even though some of the papers dealing with the use of RS do use raster GIS-style overlay procedures to define their distribution models and others do use a spatial approach but do not mention GIS).

Little is generally said about model assumptions. Of the 82 papers reviewed, only 21 discuss their assumptions. Those that do generally limit their discussion to the statistical assumptions of the technique used to perform the analysis. Very few deal with the biological and ecological assumptions and tend to take them for granted.

When dealing with ecological modeling, we need to take

into account both biological and methodological assumptions, along with some general assumptions that may limit the applicability of the results produced.

Validation, a step that is evidenced at different levels in the data flow diagram, is generally limited to the accuracy of the result of the analysis (e.g., distribution map); nothing is said about the accuracy of the original data sets (e.g., GIS data layers, observation locations) and no consideration is given to issues such as error propagation in GIS overlay.

Only 15 papers validate of the accuracy of their results based on an independent estimate of the distribution (either through comparison with an independent set of observations or through comparison with the known distribution of the species); interestingly, 50 percent of these papers are based on the deductive approach.

In fact, it should be noted that because observation data sets are the most expensive data to be collected within the general framework of setting up a GIS species distribution model, the deductive approach is the most costeffective if seen from the validation point of view.

In fact, to avoid bias, a model developed with an inductive approach cannot be validated using the same data set used to derive the species—environment relationship. Thus validation can be performed either with a second, independent data set or by dividing the original data set into two subsets, one of which is used to derive species—environment relationships and the other to validate the resulting model.

Finally, it is interesting to note that the multidimensional power of GIS is still not backed up by adequate quantity and quality of geographic data sets. This is reflected in the number of environmental variables used in analysis. In the papers reviewed, the average is just below 4.8, and only 9 out of 82 analyse more than 9 environmental variables, whereas 23 papers base their distribution models on only one environmental variable, generally vegetation.

MODELING ISSUES

Based on the results of the literature review, we have

identified five major issues that must be addressed to allow a sound GIS modeling of species distributions.

These range from uncertainties in the objectives of the research to the lack of adequate support for the assumptions underlying the implementation of GIS models. A problem that is gaining awareness is that of scale, in both time and space, but it still suffers from inadequate tools.

Slightly different is the issue of data availability, which is rarely addressable by the biologist concerned with species distribution modeling but limits the type of models that can be developed. Finally, a review of sources of errors and ways of estimating the accuracy of a GIS model addresses the problem of validation.

CLEAR OBJECTIVES

When setting up an ecological model, the very first step to be considered is clear statement of the model's objective. There is great confusion about the objectives of many published papers. This may caused by overqualification of the tool, in the sense that use of the tool becomes the objective of the paper, or by uncertainty in defining the model's goals, along with coexisting purposes of predicting or understanding.

For instance, most of the papers based on the inductive approach deal with the definition of a species—environment relationship without specifying whether they intend to analyse the relationship of cause and effect or just use the relationship as a functional description of the effect.

In the first case, the goal would be to evidence the limiting factors that are related to the species' biological needs and that drive the distribution process; in the second, it would be the simple use of correlated variables whose distribution is functional to the description of the species' distribution.

Basically, we can summarize species needs as food, shelter, and adequate reproduction sites. When using the distribution of an environmental variable to describe the species' distribution we implicitly assume that there is a correlation between these basic needs and the environmental variables used.

This correlation can be causal; that is, it describes the species' basic needs. In such cases we can identify a function that within a reasonable range of values associates each value of the environmental variable to a measure of the fulfillment of the species' basic needs (e.g., reproductive success).

But it can also be a functional description; that is, we don't really know why some ranges of values of the environmental variable are preferred by the species but we observe that the species tends to occur more frequently within those ranges. The variable might influence all the species' basic needs simultaneously or be correlated to another variable that describes one of the species' needs.

Generally speaking, the quantity and quality of the locational data and the GIS layers used in analyses are not sufficient to assess cause—effect relationships that determine the species' distribution. Furthermore, cause—effect relationships spring from the interactions of biophysical factors that range through different time and space scales; few papers take scale dependency into account in their analysis.

Moreover in this kind of analysis causal effects can be hidden by independent interfering variables or by the unaccounted stochasticity of natural events such as weather fluctuations, disturbance, and population dynamics and should be assessed in controlled environments.

We believe such uncertainties could be addressed by defining the overall goal as the assessment of the relationship that best describe the species distribution. In other words, even if the causal understanding of a relationship is not clear, whenever the species—environment relationship is able to describe the distribution of a species satisfactorily, the overall goal is achieved. Obviously the approach just described has some drawbacks. Without an adequate description of the cause—effect relationship between the species and environmental variables, models lose in transferability, in both space and time, and this limits their predictive capabilities.

ASSUMPTIONS

All models analyzed extrapolate their results to an entire

study area on the assumption of space independence of the phenomenon observed at a given place. That is, in the case of both a deductive and an inductive approach, the species—environment relationship is built on evidence that a certain species occurs somewhere and that we know the values of the environmental variables at those locations.

Obviously we know only that a species occurs at locations where it has been observed, only part of these locations have measurements of the environmental variables, and usually these measurements are collected only for the limited time range during which the investigation was carried out.

Thus, when building distribution models, evidence collected in a portion of the range is extrapolated to the entire range of occurrence of a species. In order to do so, it is assumed that the species—environment relationship used to build the model is invariant in space and time. Most of the time this is not the case, especially for species with a wide range and for generalist species. In fact, the higher the variance of the species—environment relationship, the higher the number of locations required to provide an adequate ecological profile for the species.

Second, it is generally implicitly assumed that variables that are not included in the analysis have a neutral effect on the results of the model. That is, we need to assume either that the species' ecological response to these environmental variable is constant or that the response is highly correlated with the other variables included.

Even though both of these general assumptions are very difficult to test, we believe that they should be discussed on a case-by-case basis because the result of their violation is species-specific. Errors may be negligible in certain cases but can introduce major interpretation problems in other cases.

BIOLOGICAL ASSUMPTIONS

Biological assumptions are direct consequences of the general assumptions discussed in the previous paragraph. We nevertheless believe that they are probably the most critical, but have received minimal attention in the literature.

The first assumption, which follows from the general assumption of space and time independence, states that observations reflect distribution. In other words, information on absence can be derived from observation data, which is obviously seldom the case.

In fact, any time we have a record for a species we can be sure that the species (at least occasionally) occurs at that location. In contrast, if there is no observation for a species, we can only assume that we have a record of absence if there is no bias in our sampling scheme and that we have conducted our observations over a sufficiently long period. Even then we have no way of evaluating the random effects that are intrinsic in observing animals.

These assumptions can have statistical relevance in dealing with inductive—analytical approaches, but must hold true also for the deductive models. If there is a constant bias in the visibility of a species' individuals, for instance because part of their range is less accessible than others to researchers and thus cannot be as carefully investigated, the species—environment relationship reflects this bias. For instance, observation data are often gathered through sightings carried out by volunteers, which do not follow a predefined (e.g., random) sampling scheme.

Habitat cover may limit observations to areas where the species is visible. This may create an artificial response curve that associates a positive relationship to the values of the environmental variables measured in the locations where the species is more.

Visible and a negative one in the ones measured in areas were the species has been less investigated. In such cases, we would end up mapping the areas where the species and the observers are most likely to meet, not the true distribution of the species.

This example is tailored to inductive—analytical models but can easily be extended to deductive ones, both descriptive and analytical, considering that the deductive approach is based on the a priori knowledge of specialists who rely on series of observations to gain experience and define the species—environment relationship. Again, these observations can suffer from accessibility or visibility biases.

A further assumption is that observations reflect the environmental selection of the species. Obviously this is not always true; for example, occurrences of migrant or vagrant individuals whose presence in a given location is occasional may be considered among observations.

An extreme case is represented by locust swarms blown into the middle of the desert by strong winds. Clearly, their presence does not reflect any ecological preference. Nevertheless, if we consider only the observation per se, we would conclude that high densities of locusts are found in the desert and that locusts do prefer (with all the limitations that this term carries along in such an analysis) desert environments.

Obviously the strong wind of the example should be regarded as a stochastic event and thus be treated as an outlier in the definition of a possible GIS distribution model. In other words, observations should be analyzed for their content of unconstrained selection by the species.

We will see, when dealing with the issues of scale, that GIS distribution models tend to describe only the deterministic components that drive a species' distribution pattern, so stochastic events must be either averaged on the long term or eliminated as outliers. When observations are carried out for a limited time and the biology of the species under investigation is scarcely known, this problem can become increasingly important because the identification of outliers will be virtually impossible.

STATISTICAL ASSUMPTIONS

Most of the statistical techniques used to define species—environment relationships rely on the identification of two observation sets: one that identifies locations in which the species is present and one in which it is absent. Even though this cannot be identified properly as a statistical assumption, it is probably the most important factor limiting the applicability of the statistical techniques that rely on the two groups of observations. The most common way to define the two subsets is to compare locations of known presence with a random.

Sample of locations not pertaining to the previous set. Obviously some of the random locations can represent a suitable environment for the species, thus introducing, for that particular environment, a bias that underestimates the species—environment association.

To overcome this problem, data sets can be screened for outliers, using for instance a scatter plot of the variables taken two by two. Once an outlier is identified, it can be checked to identify possible reasons for the absence of the species and, if necessary, removed from the analysis. Similar results can be achieved through analyses such as decision trees, where additional rules can be introduced to predict outliers.

Another way to get around the problem is to eliminate the absence subgroup. Skidmore et al., for example, used both the BIOCLIM approach and the supervised nonparametric classifier, which use only observation sites to derive distribution patterns.

The same result can also be achieved by using distance (or similarity) measures from the environmental characteristics of locations in which the species has been observed. A measure of distance that seems particularly promising for this application is the Mahalanobis distance.

It has many interesting properties as compared to other measures of similarity and dissimilarity, the most appealing of which is that it takes into account not only the mean values of the environmental variables measured at observation sites, but also their variance and covariance.

Thus the Mahalanobis distance reflects the fact that variables with identical means may have a different range of acceptability and eliminates the problem that the use of correlated variables can have in the analysis.

Along with the identification of presence—absence data sets, each statistical method has some specific assumption that must be satisfied for correct application of the technique. For example, nonparametric statistical tests may assume that a distribution is symmetric, whereas a parametric test may assume that the test data are normally distributed.

SPATIAL AND TEMPORAL SCALE

Scale is a central concept in developing species distribution models with GIS. This concept is common to both geography and ecology, the two main disciplines involved in the development of GIS Species distribution models.

The concept of scale evolves from the representation of the earth surface on maps and is the ratio of map distance to ground distance. Scale determines the following characteristics of a map: the amount of data or detail that can be shown, the extent of the information shown, and the degree and nature of the generalization carried out.

This group of characteristics determines the quality of the layers derived, that is, the quality of the environmental variables stored in the GIS database and the type of species—environment relationship that can be investigated using the capabilities of the GIS.

The scale of the analysis influences the type of assumptions that need to hold true for sound modeling. To clarify this concept, we need to consider that species distribution is the result of both deterministic and stochastic events.

The former tend to be described in terms of the coexistence of a series of environmental factors related to the biological requirements of the species, whereas stochastic processes are regarded as disturbances caused by unpredictable or unaccountable events. Generally distribution models are built on deterministic events and are averaged over wide spatial and temporal ranges to minimize the error related to the unaccounted stochasticity.

As we have seen, GIS distribution models rely on species—environment relationships to extrapolate distribution patterns based on the known distribution of the environmental variables. We have also seen that the relationships reflect the biological needs of the species.

The extent to which we need to coarsen our temporal and

spatial scales depends on the stochastic events that must be minimized, which in turn depend essentially on the dynamics of the species under investigation. To this extent, it is important to note that major population dynamics events happen on different scales in both time and space

As a tool for distribution modeling this graph can be of great help in defining scale thresholds toward both a minimum and a maximum scale for an analysis. For instance, when considering cause—effect species—environment relationships the processes involved (e.g., feeding behaviour) must be analyzed at an adequate scale (e.g., in our example, very detailed scale both in time and space).

On the other hand, if we need to overcome the stochasticity introduced in our observation scheme by, for instance, individual foraging behaviour we must average our results on a coarser scale in both time and space.

Thus, in GIS distribution models, both temporal and spatial scales are generally broadened so that stochastic events can average to a null component and thus be ignored. For instance, the stochasticity associated with the individual selection of a particular site, which greatly influences the distribution at a local scale, is overcome when dealing with distributions at regional scale averaging the selection of different individuals.

In a similar way, stochastic events such as local fires, which influence regional distributions when measured over a short time interval (e.g., 5–10 years), are considered outliers in an analysis that takes into account the average vegetation cover over a longer time or a wider spatial span. Similarly, we know that in short time intervals the population dynamics status of a population is highly unpredictable, whereas it may be more easily averaged on longer time scales to become scarcely predictable again at even longer intervals.

A similar consideration is intrinsic in the minimum mappable unit (MMU), a concept used largely to address spatial scale issues in GIS species distribution models that can be readily extended to the time scale. MMU can be seen from two points of view. On one hand, it is a property of the data set that is being analyzed, that is, the minimum dimension of an element (e.g., a polygon representing vegetation types of a given category, the time span between successive manifestations of a given ecological event) that can be displayed and analyzed.

On the other, it indicates the kind of averaging that must be carried out to smooth noise introduced by stochasticity. In fact, in the case of local fires, if the MMU is defined as larger than the extent of the fire in both time and space, the fire is automatically excluded from the analysis.

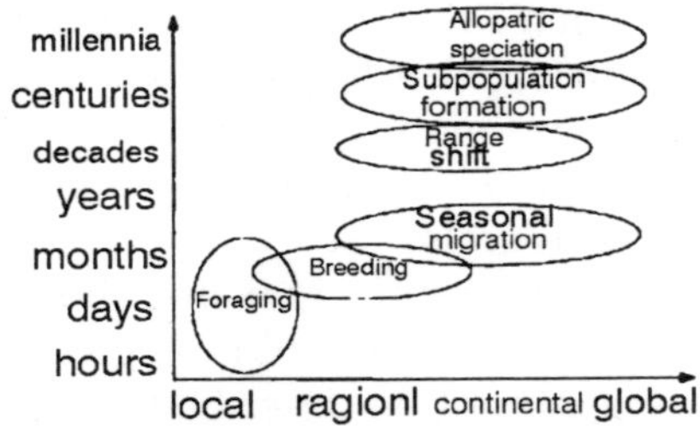

Fig. Population Dynamics Event in Relation to Time and Space Scales.

When dealing with scales on a practical basis, it should be noted that the structural complexity of distribution modeling can be simplified according to the hierarchical hypothesis that states that at any given scale particular environmental variables drive the ecological processes.

Thus weather becomes important at very broad spatial scales (e.g., continental scale). This is the basis of approaches behind models such as BIOCLIM, that of Walker, and that of Skidmore et al.; all of them describe species distribution at a continental scale in terms of their direct relationship to climatic data.

At successively finer scales such as regional landscapes, land form and topography play an important part, whereas at the most local scales, indigenous land use structures become increasingly significant to the extent that even an individual stand of timber or a single pond can play a role.

Generally speaking, the factors that are important vary according to scale, meaning that factors that are important at one scale level can lose their importance, or at least much of it, at others. As with any type of classification, the relationship between scale and environmental variables that drive ecological processes should not be taken too rigidly, and although most authors tend to agree that for broader scales climate is the most important factor, the same cannot be said when trying to identify the driving forces at finer scales.

For instance, variables considered useful at coarser scales are used in detailed studies, as in the cases of Pereira and Itami and Ross et al, which use topography to explain species distribution at a much finer scale than the regional one. The same consideration applies to the studies of Aspinall and Matthews, which use climatic data on a regional scale. On the other hand, land use is often used in distribution models developed at regional scale.

Finally, we must consider that distribution is the result of the interaction of many different biological events and that an ecological event cannot be described exhaustively on any single specific scale, but is the result of complex interactions of phenomena happening at different scales.

Thus the limit of the applicability of a given environmental variable to describe distribution on any given scale may not be so sharp and the challenge is toward the integration of different scales in the description of the species' distributions. Buckland and Elston gave an example of the integration of environmental variables stored at different resolutions within the same distribution model.

It is important to note that the concept of scale not only determines the biological extent to which a distribution model can be applied but also affects the use that can be made of such a model for conservation. Also, conservation actions can be seen as having a hierarchical approach. For instance, Scott et al. identified six different levels of intervention: landscape, ecosystem, community, species, population, and individual.

Not surprisingly, conservation actions tend to become more effective and less expensive when the assessment moves toward broader scales, that is, when one moves from the individual to the landscape approach. Obviously this relates only to the extent of the analysis, not to its resolution. Nevertheless, on a cost—benefit basis, it is generally more efficient to address conservationrelated issues at a coarser scale, which enables a landscape approach, than to concentrate on a more detailed scale (e.g., individual or population level), which requires high-resolution data to be analyzed that are either too precise or simply too abundant in terms of storage requirements to be analyzed profitably with a landscape approach.

What economics suggests is that conservation science needs to have a broader view of phenomena. A broad-scale approach and the possibility of predicting the potential dynamics of spatial patterns are needed to manage fragmentation of suitable environments and the inevitable metapopulation structure of the resulting population.

May (1994) indicates that when multiple levels of biological organization are concerned, as in a typical conservation action, the best management approach can be achieved on the regional landscape scale (103 to 105 km2). This scale level has suffered historically from limitations in the tools available for consistent analysis and is the one that has gained the most from the evolution of GIS; in fact, most of the distribution models based on GIS address problems at regional landscape level.

DATA AVAILABILITY

Data availability and quality are two of the three limiting factors in the development of GIS-based species distribution models. The problem of developing extensive data sets of environmental variables is limited by economic and political rather than technical constraints. Estes and Mooneyhan list a number of different attitudes of governments throughout the world that limit the availability of high-resolution, "sciencequality" environmental data sets.

These range from military classification of the data, thereby precluding the use of the data to the scientific community, to the low political priority that certain governments give to environmental issues. Moreover, even when policy is not an obstacle to the production and availability of data sets, entire nationwide data sets are sometimes lost during revolutions, wars, and civil disturbances.

To this it should be added that some governments (e.g., the European Union countries) ask high prices for data sets, which are generally acquired with tax money, actually preventing their broad use in any type of activity and more specifically in environmental research.

In many cases, high-quality site-specific data sets are generated for a particular research project but are compiled with nonstandard techniques, rendering them unsuitable for combination and the achievement of more extensive knowledge of an area.

In the past few years there has been an increasing effort to develop metadatabases of available data sets throughout the world, and the problem is being addressed by national and international organizations (e.g., United Nations Environmental Programme, World Bank, U.S. Geological Survey [USGS], European Environmental Agency). These initiatives still do not address the problem of producing high-quality data sets, but at least they are a start in collating existing data sets.

An important example is given by the joint efforts of the USGS, the University of Nebraska—Lincoln, and the European Commission's Directorate General Joint Research Centre, which are generating a 1km-resolution Global Land Cover Characterisation (GLCC) database suitable for use in a wide range of environmental research and modeling applications from regional up to continental scale. All data used or generated during the course of the project (source, interpretations, attributes, and derived data), unless protected by copyrights or trade secret agreements, are distributed through the Internet.

This effort goes in the direction of producing and distributing homogeneous medium-resolution high-quality data sets with known standards of accuracy. Further aspects of raw data sets are discussed in the next section, where the quality of the data used to build models is discussed. We do not discuss this issue further here because we do not believe it to be a problem that can be addressed directly by conservation biologists or ecologists, although they can contribute to developing awareness of the need for standardization of data sets and for their production and dissemination.

VALIDATION AND ACCURACY ASSESSMENT

Generally, the main function of a GIS-based species distribution model is to produce a map or its digital analogue for assessment of management and conservation actions. Possibly the most important question to be asked by a user is 'how accurate is the distribution map that has been produced?'

Many articles have been written on the sources of error in the data layers that may be included in a GIS. Nevertheless, few authors of papers dealing with animal distribution include an assessment of the accuracy of their model and a validation of the product.

Because we believe this issue to be central to the entire process of species distribution modeling, the aim of this section is to review sources of error in GISS, to discuss methods of assessing mapping accuracy, and to evaluate the accumulation of thematic map errors in GISS, thus providing a framework for assessment of the accuracy of distribution models developed with GIS.

Source of Errors

GIS data layers are traditionally classified according to their data structure, either raster or vector. To a certain extent, both error sources and accuracy evaluation methods have been investigated following this traditional classification.

Raster images may be obtained from remote sensing instruments carried by aircraft or spacecraft platforms, or by converting an existing line map (vector data structure) to a raster data structure. Two types of error are inherent in remotely sensed images: geometric and radiometric. These error sources are addressed in detail in numerous monographs and papers, including Colwell and Richards.

A raster image is usually made up of a regular grid of adjacent rectangular cells or pixels (i.e., a rectangular tessellation). Geometric error in a remotely sensed image is caused by movement in the remote sensing platform; distortion caused by the earth's curvature and terrain; different centrifugal forces from earth.

Affecting spacecraft movement; the earth's rotational skew; distortions introduced by the remote sensing device itself, including systematic distortions caused by sampling sequentially from each detector and nonlinear scanning; and errors introduced by the georeferencing process. Geometric error causes a point on the remotely sensed image to occur in the wrong position relative to other points in the image.

Correction of geometric errors in remotely sensed data is now a routine aspect of their preprocessing. The map or image is usually rubber-sheeted to fit it to an appropriate map projection. Corrected images with geometric errors of less than 0.5 pixel are now obtainable and acceptable.

However, the base maps from which control point information is derived may be of poor quality. Bell reported that maps used to geometrically correct images of the Great Barrier Reef contained errors of up to 1 km. The accurate selection of control points is crucial in obtaining acceptable results.

Points within a rubber-sheeted image are no longer on a regular grid because they have been warped to fit into the projection defined by the ground control points (GCPS). To obtain a regular grid, an interpolation method is used to nominate a value for a regular grid point that falls between the points in the rubber-sheeted image. Lam provides an excellent review of other interpolation methods, including splines, finite difference, and kriging.

Radiometric errors occur as a result of differential scattering of electromagnetic radiation at varying wavelengths, sensors that have poorly calibrated multiple detectors within a band, sensor calibration error, signal digitization error, and scene-specific error such as off-nadir viewing, irradiance variation, and terrain topography (Richards 1986).

Correction of band-to-band distortion is performed using image histograms (shifted to the origin to remove atmospheric scattering effects), whereas line striping effects are reduced by calibration of detectors or by matching detector statistics during computer processing.

A final type of error may be caused by a time lag between ground truthing and image collection. In this case, pixels may be noted as incorrect in the error matrix when they may be actually correct at the time of image acquisition. Vector images have been traditionally recorded and stored as maps. Maps are subject to many errors. Some errors are introduced during the creation of the map, such as the original line smoothing, which may not follow the true isolines on the ground. Other errors may be associated with the physical medium used to store the map (e.g., paper stretch and distortion).

Maps may be represented in computer GISS by a variation of the vector data structure or converted to a raster data structure. In its simplest form, the vector data structure has map lines approximated to a set of points (nodes), which are linked by lines (or arcs). Vector data may be obtained by digitization.

Digitization introduces a number of errors. Varying line thickness on the original map requires automatically scanned vector lines to be thinned. During manual digitization the centre of the map line must be followed carefully if the map lines vary in thickness. This requires very careful hand digitizing or high-accuracy automatic scanners.

The number of vertices (points) used to approximate a curve is also critical. Too few vertices will result in the line appearing stepped, and too many vertices create large data volumes. Thus, even with extreme care, error is introduced during digitization. As for raster images, the main method of correcting geometric error in vector images is by using ground control points from a cartographically correct map to transform the vector image to a known projection.

DATA LAYER ERROR QUANTIFICATION

Methods for quantifying error in a raster data layer are

based on the error matrix (also called a contingency table or confusion matrix) concept, first expounded for remotely sensed data in the 1970s.

The aim of the error matrix is to estimate the mapping accuracy (i.e., the proportion of correctly mapped pixels) within an image. An error matrix is constructed from points sampled from the image.

The reference (or verification) data are normally represented along the columns of the matrix, and are compared with the classified (or image) data represented along the rows. The major diagonal of the matrix represents the agreement between the two data sets.

To check every cell for correctness would be impossible except in the smallest map area, so various sampling schemes have been proposed to select pixels to test. The design of the sampling strategy, the number of samples required, and the area of the samples have been debated by the remote sensing and GIS community.

As with any sampling problem, one is obviously trying to select the sampling design that gives the smallest variance and highest precision for a given cost. A number of alternative designs have been proposed for sampling the pixels to be used in constructing the error matrix. Berry and Baker recommended the use of a stratified systematic sample.

The advantage of systematic sampling over random sampling is that sample units are distributed equitably over the area. The disadvantage is that the resulting sample is weighted in favour of the class covering the largest area and that classes with a small area may not be sampled at all.

Simple random sampling in land evaluation surveys emphasizes larger areas and undersamples smaller areas. Zonneveld suggested that a stratified random sample was preferable, and Van Genderen et al. (1978:1135) agreed that a stratified random sample is "the most appropriate method of sampling in resource studies using remotely sensed data."

Rosenfield et al. suggested a stratified systematic unaligned sampling procedure (i.e., an area weighted procedure) as a first-stage sample to assist in identifying categories occupying a small area, followed by further stratified random sampling for classes with fewer than the desired minimum number of points. Todd et al. argued that singlestage cluster sampling is the cheapest sampling method because multiple observations can be checked at each sample unit on the ground.

Congalton simulated five sampling strategies (simple random sampling, stratified random sampling, cluster sampling, systematic sampling, and stratified systematic unaligned sampling) using a different number of samples over remotely sensed images of forest, rangeland, and grassland.

The aim of the study was to ascertain the effect of different sampling schemes on estimating map accuracies using error matrices. He concluded that great care should be taken in using systematic sampling and stratified systematic unaligned sampling because these methods could overestimate population parameters.

Congalton also stated that cluster sampling may be used, provided a small number of pixels per cluster are selected (he suggested a maximum of 10 sample pixels per cluster). Stratified random sampling worked well and may be used where small but important areas must be included in the sample. However, simple random sampling may be used in all situations.

The number of samples may be related to two factors in map accuracy assessment: the number of samples that must be taken in order to reject a map as being inaccurate and the number of samples required to determine the true accuracy within some error bounds for a map.

Van Genderen et al. pointed out that we want to know, for a given number of sample pixels, the probability of accepting an incorrect map. In other words, when high mapping accuracy is obtained with a small sample (e.g., 10 items), there is a chance that no pixels that are in error may be sampled (i.e., a type II error is committed). The corollary, as stated by Ginevan, is also important: The probability of rejecting a correct map (i.e., committing a type I error) must also be determined.

Van Genderen and Lock and Van Genderen et al. argued that only maps with 95 percent confidence intervals (i.e., b = 0.05) should be accepted and proposed a sample size of 30. Ginevan pointed out that Van Genderen et al. made no allowance for incorrectly rejecting an accurate map.

The tradeoff one makes using Ginevan's more conservative approach is to take a larger sample, but in so doing reduce the chance of rejecting an acceptable map. Hay (1979) concluded that minimum sample size should be 50, greater than that of Van Genderen and Lock.

The number of samples must be traded off against the area covered by a sample unit, given a certain quantity of money to perform a sampling operation. Curran and Williamson asked whether many small-area samples or a few large-area samples should be taken.

The answer is that it depends on the cover type being mapped; a highly variable cover type such as rainforest is better suited to many small-area samples, whereas for more homogeneous cover types it is more efficient to take fewer large-area samples.

Generally, mapping of heterogeneous classes such as forest and residential land is more accurate at 80-m resolution than at finer resolutions such as 30 m; however, more homogeneous classes such as agricultural land and rangeland are more accurately mapped at 30 m than at 80 m. The reason for this is the tradeoff between ground element size and image pixel resolution.

Based on the error matrix, different measures of accuracy can be derived. A commonly cited measure of mapping accuracy is the overall accuracy, which is the number of correctly classified pixels (i.e., the sum of the major diagonal cells in the error matrix) divided by the total number of pixels checked. Anderson et al. suggested that the minimum level of interpretation accuracy in the identification of land use and land cover categories should be 85 percent.

Overall classification accuracy is the ratio of the total number of correctly classified pixels to the total number of pixels in each class.

Cohen and Bishop et al. defined a measure of overall agreement between image data and the reference (ground truth) data called Kappa or K. K ranges in value from 0 (no association, that is, any agreement between the two images equals chance agreement) through 1 (full association, or perfect agreement between the two images). K can also be negative, which signifies a less than chance agreement.

The methods just discussed for quantifying error in raster images are equally applicable to quantifying error in vector polygons. Instead of checking whether an image pixel is correctly classified, a point within the polygon is verified against the ground truth information.

A specific problem encountered with vector images is ground truth samples that occur across boundary lines; in this case the class with the largest area within the sample area may be selected to represent the vector map image.

A method of assessing map accuracy based on line intersect sampling was described by Skidmore and Turner. Line intersect sampling is used to estimate the length of cover class boundaries on a map that coincide with the true boundaries of the cover classes on the ground. A ratio of coincident boundary to total boundary is proposed as a measure of map accuracy and this ratio is called the boundary error. Although this technique was developed for vector maps, it is equally applicable to raster maps.

Skidmore and Turner found that the true boundary lengths were not significantly different from the estimated boundary lengths sampled using line intersect sampling, with a2 = 0.05. The estimated boundary accuracy (64 percent) was extremely close to the true boundary accuracy (65.1 percent), and there was no significant difference between the true and estimated boundary accuracy.

RELIABILITY OF THE OUTPUT

Apart from assessing the accuracy of the original data sets used to produce GIS models, the final products of the modeling effort must be validated.

As previously evidenced, the process of model building

takes its move from a number of data maps of the same region (e.g. elevation, soils, ground cover), which are digitized and geographically rectified to a common projection in a GIS. Such maps may be stored as a series of layers in a GIS.

Each point within a polygon, or each cell in a raster layer (where each raster cell is assigned one value), takes the values of the layers directly above the point. The model is then built defining specific questions that may be asked about specific points or cells, such as the slope or aspect at the point.

Eventually the biologist is interested in mapping areas that satisfy the known ecological requirements of a species (e.g., slope greater than 30°, with erodible soils, occurring on a southerly aspect). By overlaying the map layers, such simple queries may be answered.

Nevertheless, the simple process of overlaying layers propagates the errors of the original data and sometimes amplifies them. If the sources of errors are identified and their accuracy is known, an estimate of the accuracy of the final output can be achieved through error propagation analysis.

Sources of errors range from those evidenced in the previous paragraphs to those inherent to nonspatial or text data that are also part of a GIS model. For instance, nonspatial data include knowledge or rules used by expert systems. All these errors contribute to error accumulation when overlaying GIS data layers.

Composite overlaying is the simplest overlaying technique. Two or more layers are combined, and the raster locations (or polygons formed) describe the union of the classes on the layers. The composite overlay is in effect a universal Boolean and operation over the whole map.

That is, for a two-layer data set comprising layer X and layer Y, we note X i=1, n Y j=1,m (i.e., the intersection of X and Y for the n classes in map X and the m classes in map Y) at all points over the map.

Arithmetic and mathematical operators that may be applied to two or more layers include addition, subtraction, multiplication, division, maximum, minimum, average, and exponent. The method by which error is accumulated during the overlaying process is important for modeling error in the final map products. The first necessity for modeling map error accumulation is to quantify the error in the individual layers being overlaid. A lot of work has been done on this problem, with some tangible results.

Newcomer and Szajgin used probability theory to calculate error accumulation through two map layers. They assumed that the two map layers were dependent; that is, if we select a cell that is in error in layer 1 then that act reduces the probability of selecting an erroneous cell from layer 2. If the data layers are independent, then an erroneous cell selected from layer 1 does not reduce the probability of selecting an incorrect cell from layer 2.

Using the statistics of Parratt with empirical data, Burrough concluded that with two layers of continuous data, the addition operation is unimportant in terms of error accumulation. The amount of error accumulated by the division and multiplication operations is much larger.

The largest error accumulation occurs during subtraction operations. Correlated variables may have higher error accumulation rates than noncorrelated data because erroneous regions tend to coincide and concentrate error rates there.

The use of Bayesian logic for GIS overlaying is explained in Skidmore. As with Boolean, arithmetic, and composite overlaying, there is inherent error in the individual data layers when overlaying using Bayesian logic. In addition, Bayesian overlaying uses rules to link the evidence to the hypotheses; the rules have an associated uncertainty and are an additional source of error.

A number of possible solutions are suggested for modeling accumulation of error in a GIS. These include reliability diagrams and a probabilistic approach.

Wright suggested that reliability diagrams should accompany all maps. He also emphasized that the sources used to generate different regions of the map have varying accuracy and these sources should be stated clearly on the map. For example, one region may have been mapped using low-altitude aerial photography and controlled ground survey, and would therefore be more accurate than another region mapped using high-altitude photography and only reconnaissance survey.

This theme was taken up by Chrisman and MacEachren, who suggested that such a reliability diagram showing map pedigree should be included as an additional layer accompanying each map layer in a GIS. However, for the purposes of error accumulation modeling, reliability diagrams do not provide a quantitative statement about the accuracy (or error) of the map.

The supervised nonparametric classifier described by Skidmore and Turner and Skidmore et al. classified remotely sensed and GIS digital data. The classifier gives for all cells the empirical probability of correct classification for each class according to the training area data and thereby gives an indication of map accuracy.

The few methods proposed for modeling error accumulation are limited in their application. Working with ideal data, these methods do allow some conclusions to be drawn about error accumulation during GIS overlay operations. However, the methods break down when used with map layers created under different conditions than assumed by the methods.

Newcomer and Sjazgin used probability theory to model error accumulation, whereas Heuvelink and Burrough modeled the accumulation of error in Boolean models, using surfaces interpolated by kriging as the estimated error source.

SENSITIVITY ANALYSIS

When no information is available on the extent of the errors of the original data sets or on the type of error propagation function applicable to the model, a way of defining levels of reliability of the output is to analyse its variability subject to changes in the input parameters. We have seen possible sources of variability and uncertainties that can arise in the deductive and inductive approaches in species distribution modeling.

They range from subjective errors introduced by the

specialist who defines the species—environment relationship to locational errors of species observation caused by possible biases in the sampling scheme or inaccuracy of the instrument used to locate the species (e.g., radiotelemetry, global positioning system). A through discussion of the sources of uncertainties in species distribution modeling can be found in Stoms et al..

Once the sources and ranges of variability are identified, different input data sets can be systematically produced, selecting the variates from the variability range of the original input variables. These alternative data sets are used to build alternative models that can be compared with the original one, identifying the variability induced in the output by the uncertainty of the input variables.

The variability induced in the output is a measure of the overall performance of the model and can be compared with a predetermined acceptable significance level. As a general rule, when dealing with great uncertainties in the measures of the input variables, a greater inertia (less subject to changes in the results) of the model is generally preferable.

Sensitivity analysis does not replace validation but can be used at any stage of the model-building process to identify the parameters that should be monitored more carefully to maximize the reliability and the accuracy of the results. The use of GIS in species distribution modeling should follow precise steps. First of all, we recommend more unambiguous use of some key terms such as scale and habitat.

The latter seems to be a particular problem not limited to GIS applications but spanning the entire field of ecological studies. We believe that GIS can be a valid tool to overcome the current ambiguity between the species-related and land-related concepts of the term habitat.

As a matter of fact, the latter was introduced as a way of dealing with problems related to environmental mapping using traditional tools, and the enhancements introduced by GIS do overcome those problems. In the meantime, however, we suggest replacing the word habitat with more unproblematic terms such as environment.

The ambiguity of the term habitat and most of the works on habitat use and habitat selection have also given rise to the question of whether GIS-based models can be used to explain the causal event of a species—environment relationship. In our opinion, use of GIS is not central to a better understanding of causal effects in a species—environment relationship, especially if the quality of data does not support both high-resolution and large-extension analyses.

Currently our analytical capabilities are limited by the lack of high-resolution, global coverage data sets. Nevertheless, the use of GIS's spatial analysis tools in the framework of a controlled environment in which all the key variables are monitored at an adequate resolution can increase our ability to assess causal effects in species—environment relationships.

Apart from further generic considerations, we think that a few important issues have been overlooked in these first years of application of GIS to the field of ecological modeling and especially in the field of species distribution modeling.

There has been inadequate discussion and consideration of the assumptions underlying the model-building process and the related issues of spatial and temporal scale, which are of paramount importance for sound scientific use of GIS. Adequately discussed assumptions can justify the development of a model.

Whenever a hypothesis is stated and a model is built to test its congruence, it should be regarded as a problem-solving tool. For instance, we will never really know what will be the outcome of alternative management options, but we can state different hypotheses, state the assumptions that must be met to make each hypothesis hold true, and try to model the result of the different options. In such cases we don't have direct control over the results of the management action; we can only ensure that the assumptions are met.

This means that the output of the model will hold true if the assumptions are met and if the model is built on the logical consequences of these assumptions. In such cases validation, meant as an independent estimate of the truth, can to a certain extent be neglected. Nevertheless, most of the time assumptions are not adequately discussed and this is particularly evident when dealing with the constraints of scale dependency of biological events.

Probably the issues of scale still suffer from inadequate support from the available tools. For instance, we still lack convenient ways of handling spatiotemporal data in GIS software packages, not to speak of analyzing the two components together.

If validation can be neglected somewhat when dealing with hypothesistesting models, it becomes a fundamental issue when building analytical models, which are built to assess species—environment relationships and ecological processes.

In such cases, validation steps must be included from the beginning of the model building process, first assessing the quality and reliability of the raw data used, then evaluating the limits of the relationships that drive the process and finally analyzing the correspondence of the output with the truth.

Validation can be a costly exercise in model building, and efforts are being made to find a cost-effective approach to this issue. Because the issue of validation is general to GIS modeling and especially GIS ecological applications, it can benefit from the experience derived from other fields of science (e.g., RS).

Methods for summarizing the accuracy of raster and vector maps using point samples and error matrices are now widely used in GIS and are beginning to find their way into ecological applications as well. However, standard techniques have not had universal acceptance for a number of reasons. For example, a number of alternative sampling designs have been proposed for analyzing the accuracy of imagery.

The choice of sampling design is often subject to the particular problems associated with the area to be ground-truthed. However, a number of general trends are obvious from the GIS and RS literature. Random and stratified random sampling are acknowledged to maximize precision and accuracy (though at a higher cost than cluster sampling or systematic sampling).

It should be noted that in highly heterogeneous

landscapes (e.g., native forests, especially tropical forests), stratification is often too costly to consider. Cluster sampling offers reduced sampling costs, but in order to be effective it depends on low intracluster variance.

Systematic sampling schemes may lead to a bias in parameter estimation if periodic errors align with the sampling frame (e.g., as a result of image banding or linear topographic features, as in the Allegheny Mountains of Pennsylvania).

GIS data layers contain numerous errors. These pose a number of problems as errors accumulate during the process of analysis and model building. Although modeling the accumulation of error during GIS overlay analysis is still in its infancy, some methods for measuring error accumulation during GIS analysis have been discussed.

Any procedure to reduce mapping error in individual layers in a GIS will improve the mapping accuracy of an overlay generated from the GIS. Until better error-modeling techniques are developed for GISS, descriptive statistics should ideally be calculated for each layer in a GIS, as well as for each layer produced by GIS modeling.

The descriptive statistics should include overall mapping accuracies as well as class mapping accuracies. An alternative way of defining the performance of a GIS model, thus assigning a level of reliability to its results, is sensitivity analysis that identifies crucial parameters.

These parameters are those that, within their range of variation, determine the highest variation in the model output.

Index

A

Abundance functions 150
Abundance relationships 151, 153
Ad libitum sampling 247, 251, 252
Advantages and problems 101
Analysis of data on mortality 137
Analysis of observational 253
Analysis of time series 134
Animal home ranges 41
Applications 105
Assessing habitat selection 72, 74, 91
Assumptions 20, 30, 40, 65, 85, 96, 100, 102, 111, 129, 130, 140, 151, 170, 204, 234, 253, 297, 298, 306
Availability versus 123
Available for choosing 19

B

Bands and collars 10
Basic population model 195
Behavioral parameters 211
Beware teleology 224
Biological assumptions 283, 289
Biology of vertebrates 1
Bootstrap approach 193, 194, 200

C

Cafeteria experiments 123
Choosing markers 17, 19
Circle and ellipse approaches 39
Classifications of behavioral interactions 227
Components of a pva 184
Conservation applications 213
Conventional approaches 113
Current guidance available 17

D

Data availability 291
Data layer error quantification 295
Defining habitats 74, 75
Definition of home range 27
Definition of home range 61
Demographic response design 69, 73, 101
Demographic variation 175, 176, 178, 179, 180, 184, 186, 194, 195, 203, 205, 207

Dependence 78, 79, 133, 134, 135, 136, 137, 138, 139, 140, 141, 142, 143, 147, 184, 195, 196, 197, 198, 199, 202, 215, 251, 283, 284
Detection of causes of 143
Detection of density 134
Dominance 212, 216, 227, 228, 229, 230, 232, 233, 234, 235,
Dynamic interactions 58

E

Effects of markers among 4
Evaluating the importance 122
Evolution of sociality 212
Exclosures 117, 118, 129
Experimental manipulation 125, 145, 147, 148

F

Feces or pellets 118
Feeding site surveys 114
Focal sampling 248, 249, 251
Fourier series 39
Fractal estimators 41

G

Gastrointestinal samples 121
General structure of gis-based models 274

H

Habitat definitions and use 270
Habitat evaluation in delusions 65
Harmonic mean distribution 40
Home range core 51, 55
How to describe 211, 215

I

Identifying the individual 245
Improvements on lead animal studies 124
Improving index surveys 155
Improving sample resolution 130
Incorporation of parameter 199
Indirect estimation 192
Individual variation 181, 182, 183, 184, 192, 206, 228
Inferring habitat quality 96
Information obtained 4, 120, 130, 131
Innovations 124, 129, 211
Instantaneous scan sampling 250

K

Kernel estimators 46, 47, 48, 49, 50, 51, 61, 63
Key factor analysis 143, 144, 145

L

Lead animals 114, 124
Lessons 61, 183, 212

M

Mammalian societies 211, 216, 244
Mammals 13, 14, 16, 18, 19, 23, 25, 28, 31, 55, 69, 73, 74, 104, 107, 118, 119, 120, 130, 149, 151, 162, 163, 170, 198, 211, 220, 229, 253
Marker evaluation studies 1, 4, 14, 15, 16, 24, 25, 26
Measuring habitat availability

79, 80
Measuring habitat use 75
Methods for behavioral 245
Methods for evaluating 69
Minimum convex polygon 38, 39, 43, 61
Modeling issues 280
Monitoring indices over time 159
Monitoring populations 149
Monitoring programs 150, 156, 157, 160, 161, 162, 163, 164, 165
Mutilation 6, 8, 13, 14, 17, 18, 22

P

Population change 134, 145, 151, 152, 153, 157, 160, 166, 168
Population viability analysis 167, 185, 199
Postingestion samples 118, 120
Power estimation 160
Predictability of behaviour 262
Process variation 183
Process variation 172, 183, 187, 207
Protocols and technological advances 22

Q

Qualitative observations 169
Quantifying home range 55

R

Radiotransmitters 6, 9, 11, 12, 14, 18, 23, 25, 124
Recent ecological studies 20
Reliability of the output 299, 302
Reptiles and amphibians 7, 8, 9, 17, 18

S

Sample resolution 129, 130, 131
Sampling and recording rules 247, 252
Sampling requirements 164
Searching for a behavioral 261
Sensitivity analysis 302, 303, 306
Sequences through the mist 263
Setting objectives 165
Site attribute design 72, 73, 101
Social dynamics 209, 210, 211, 212, 213, 115, 216, 217, 218, 244, 263
Source of errors 293
Sources of variation affecting 172
Spatial and temporal scale 287
Spatial aspects of measuring changes in indices 157
Spatial variation 126, 180, 182, 185, 198, 200, 206
Static interactions 56
Stationarity 222, 228, 235
Statistical assumptions 279, 285
Statistical rationality 254
Stochastic variation 173
Survey of food remains 115

T

Tagging 4, 5, 6, 7, 8, 13, 17, 21, 22
Tagging and mutilation 17
Techniques for behavioral 251
Temporal variation 163, 171, 177, 179, 184, 186, 187, 193, 194, 203, 205, 206, 207
Terminology 66, 215, 269
Terrestrial vertebrates 112, 118

Territories 32, 33, 34, 37, 104, 111
Testing for territoriality 59
The ethogram 222, 262
The role of foraging theory 127
Time sampling 247, 249
Types of indices 150

U

Uncertainty into persistence 199
Understanding ourselves 213
Use of isotope ratios 124

V

Variability of index 153
Variability of indices of animal 161

W

Which markers to use 3
Why study social dynamics 212